LONDON MATHEMATICAL SOCIETY LECTURE NOTE SERIES

Managing Editor: Professor J.W.S. Cassels, Department of Pure Mathematics and Mathematical Statistics, University of Cambridge, 16 Mill Lane, Cambridge CB2 1SB, England

The books in the series listed below are available from booksellers, or, in case of difficulty, from Cambridge University Press.

London Mathematical Society Lecture Note Series. 114

Lectures on Bochner-Riesz Means

Katherine Michelle Davis
The University of Texas at Austin

Yang-Chun Chang
*The University of Texas at Austin and
Beijing Normal University*

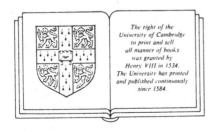

The right of the
University of Cambridge
to print and sell
all manner of books
was granted by
Henry VIII in 1534.
The University has printed
and published continuously
since 1584.

CAMBRIDGE UNIVERSITY PRESS

Cambridge

New York   New Rochelle   Melbourne   Sydney

Published by the Press Syndicate of the University of Cambridge
The Pitt Building, Trumpington Street, Cambridge CB2 1RP
32 East 57th Street, New York, NY 10022, USA
10, Stamford Road, Oakleigh, Melbourne 3166, Australia

First published 1987

Library of Congress cataloging in publication data available

British Library cataloguing in publication data

Davis, Katherine Michelle
Lectures on Bochner-Riesz means. – (London Mathematical
Society lecture note series, ISSN 0076-0552; 114)
1. Fourier series   2. Convergence
3. Bochner-Riesz means
I. Title   II. Chang, Yang-Chun   III. Series
515'.2433   QA404

ISBN   0 521 31277 9

To John,

Who stood by through the changes.

CONTENTS

INTRODUCTION

The purpose of this book is to give a self-contained exposition of the geometric theory of Bochner-Riesz means. The subject deals with the most basic topic in Fourier analysis, the question of when a Fourier series converges to its original function. Substantial progress was made in the mid 1970's, but the techniques are still avaliable only in the technical literature. Our intent is to present an account accessible to graduate students. We have slighted certain important topics in order to maintain a consistent presentation. We have assumed that the reader is familiar with real analysis at a graduate level, and with basic facts about distributions and the Fourier transform. A basic reference is the text by Stein and Weiss, Introduction to Fourier Analysis on Euclidean Spaces [50] , and the texts of Rudin, [43] and [43].

In writing this book, we benefitted with extensive conversations over many years with our colleagues. We wish to thank Professors E. Fabes, R. Fefferman, and E. M. Stein for their help with the material in Chapters 1, 2 and 3. The contents of Chapters 4 and 5 were influenced by conversations with Professor A. W. Knapp. For the general philosophy of Chapters 7 and 8 we are indebted to Professors A. Cordoba and C. Fefferman. The first draft of this book was written while the first author was supported by NSF grants MCS 8202165 and 8001799. The first author is indebted to Professor W. Beiglbock and the Institut fur Angewandte Mathematik at Heidelberg for the opportunity to finish the penultimate draft of this book. The second author wishes to thank the Ministry of Education of the People's Republic of China for their support during the entire project. Both authors wish to thank Professor K. Merryfield for his critical reading of an early version of this work.

Both authors are grateful to the University Research Institute for financial support during the production of this book.

Finally, the first author thanks Margaret Combs and Jan Duffy, who were always willing to help.

CHAPTER 0

0.1 INTRODUCTION

In this section we give a quick discussion of the main topic of the book: when does a Fourier series converge? We let T^n denote the n-torus $[0,1)^n$, with elements $\theta = (\theta_1, \theta_2 \ldots, \theta_n) \in T^n$, and $k = (k_1, k_2, \ldots, k_n)$ an n-tuple of integers. If f is in $L^p(T^n)$, then the Fourier transform and the formal Fourier series associated to f are defined as

$$\hat{f}(k) = \int f(\theta)e^{-2\pi ik\theta}d\theta$$

$$f = \sum_{k \in \mathbb{Z}^n} \hat{f}(k)e^{2\pi ik\theta}.$$

The central topic of this book is an analysis of the sense in which the formal Fourier series of f actually converges to f. Of course we have to take finite sums, and then a limit. Since the sum is over all n-tuples of integers, that is, over all $k \in \mathbb{Z}^n$, we need some method to order the lattice points of $\mathbb{Z}^n$. The simplest is to include them all in ever-expanding spheres; define $|k|^2 = k_1^2 + k_2^2 \ldots + k_n^2$, and then define the Rth spherical partial sum of the Fourier series of f as

$$S_R f(\theta) = \sum_{|k| < R} \hat{f}(k)e^{2\pi ik\theta}.$$

With this definition, the basic question we want to study is: when does $S_R f$ converge to f in L^p?

In one dimension, the answer is classical, and was found by M. Riesz in 1910. (See Zygmund [57], Chapter 7, 2.4). Convergence is valid for all f in L^p if and only if $1 < p < \infty$.

In higher dimensions, the question was open until very recently. Carl Herz showed in 1954 that a necessary condition for convergence is that $\frac{2n}{n+1} < p < \frac{2n}{n-1}$. Charles Fefferman showed in 1972 that convergence in L^p holds if and only if $p = 2$. Fefferman's proof and its consequences are the main subject of this book.

Fefferman's result shows that this is not the right way to sum Fourier series. What other ways are there? Well, the points of $\mathbb{Z}^n$ could be grouped in some order other than taking whatever is inside a sphere. Concentric polygons are an obvious thing to try, but this turns out to be no more interesting than repeating several one-dimensional results. It doesn't give any new mathematics, and it avoids having to think deeply about Fefferman's result. To avoid thinking about a subject is almost always a mistake; at best you are in for some big suprises later on.

A classical analyst would quickly tell you the alternative to using spherical partial sums: use a summability method. We shall analyse a method introduced by S. Bochner, which itself was a variation of a summability method of Riesz: Set $(\xi)_+ = \xi$ if $\xi > 0$; let it be 0 if $\xi < 0$. Then we define Bochner-Riesz means of order α by

$$S_R^\alpha f(\theta) = \sum_{|k| < R} \left[\left(1 - \frac{|k|^2}{R^2} \right)_+ \right]^\alpha \hat{f}(k)e^{2\pi ik\theta}.$$

The point here is that if $\alpha = 0$, $S_R^\alpha f = S_R f$, so that by studying the limiting behaviour as α tends to zero, we can hope to understand what's wrong with S_R.

Very well; what is known about S_R^α for α near zero? The behaviour turns out to be very complicated in high dimensions, and in fact the complete answer is known only in two dimensions. Instead of trying to write down a formula for the answer, we'll draw a picture. Figures 1 and 2 below show the L^p boundedness of S_R^α for $1 \leq p \leq \infty$. The vertical axis is indexed by α; the horizontal by $\frac{1}{p}$. Dotted lines and open circles represent points of known unboundedness; shaded regions known boundedness.

The purpose of this book is to give detailed proofs of the results in the pictures.

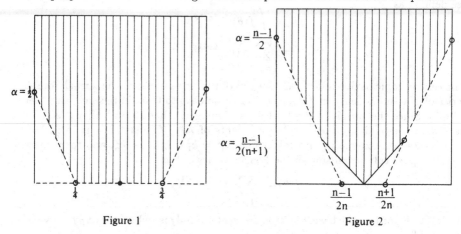

Figure 1   Figure 2

0.2 THE ROLE OF FUNCTIONAL ANALYSIS

In this section we give an idea how convergence questions can be attacked. The best way to do this is through a fast review of a standard result: there is a continuous function in T whose Fourier series diverges at $\theta = 0$. Of course this function can be explcitly constructed, but in general dimension explicit computations are unmanageable. We turn for relief to the methods of functional analysis.

The abstract problem is this. If I take a finite Fourier series, the partial sums of its Fourier series simply stop after a while, and I get the original function back. So, partial sums of Fourier series converge on the so-called trigonometric polynomials. The trig polynomials are dense in the continuous functions (Stone-Weierstrass Theorem) and the continuous functions are dense in L^p. I need to pass from information about convergence of partial sums on a dense subset to convergence on a whole space of functions: I need to interchange limits. Standard real analysis tells me that some sort of uniform convergence allows interchange of limits; functional analysis tells me that uniformity is a necessary condition. This is the point of the uniform boundedness theorem.

0.1 THEOREM. *There is an $f \in C(\mathbf{T})$ such that* $\sup_N |S_N f(0)| = \infty$.

PROOF: We begin by changing this into a problem about operators on function spaces. Define the operators $S_N : C(\mathbf{T}) \to \mathbb{C}$ by $S_N f = \sum_{|k| \leq N} \hat{f}(k)$. Since the

2

sum is finite, each S_N is a countinous linear functional on $C(\mathbf{T})$. Every such linear functional is given by integration against a finite Borel measure, in this case,

$$\sum_{|k| \leq N} \hat{f}(k) = \sum \int e^{-2\pi ik\theta} f(\theta)d\theta = \int K_N(\theta)f(\theta)d\theta;$$

here

$$K_N(\theta) = \sum_{|k| \leq N} e^{-2\pi ik\theta} = \frac{\sin 2\pi(N + \frac{1}{2})\theta}{\sin(\pi\theta)}.$$

So in this case, the measure giving S_N is absolutely continuous, and its total variation norm is just $\|K_N\|_1$. This is not even very difficult to compute; in Zygmund [57] the precise computation is given as

$$\|K_N\|_1 = \frac{2}{\pi}\log N + O(1).$$

The real problem is to relate how S_N acts on individual functions with how it behaves as an operator on the space $C(\mathbf{T})$. This is what the uniform boundedness theorem tells us. Either: i) The S_N are uniformly bounded in operator norm; or, ii) $\sup_N |S_N f| = \infty$ for a dense set of f. Since the first conclusion does not hold, the second does.

This result is absolutely typical of what we will do in the rest of the book. There will be some functional analysis trickery that reduces convergence questions to questions on the boundedness of operators on function spaces. The functional analysis is followed by a computation of specific operators; and the analysis concludes with a detailed computation.

0.3 BACKGROUND

In the next chapter, we will present the functional analysis needed to analyze convergence of Fourier series. The best trick will be to transfer problems from Fourier series to Fourier integrals. This is good because it is easier to compute an integral explicitly than to sum a series in a closed form. On the other hand, this is bad because the integrals defining Fourier transforms do not converge absolutely. The modern solution to this difficulty is to use a dense subset on which everything does converge, and then pass to a limit. This is the point of the theory of rapidly decreasing functions, which we summarize in this section. A more detailed treatment is given in Stein and Weiss [50].

For $f \in L^1(\mathbb{R}^n)$, the Fourier transform $\hat{f}$ is defined as

$$\hat{f}(\xi) = \int f(x)e^{-2\pi ix\xi}dx.$$

The integral converges absolutely, and $\|\hat{f}\|_\infty \leq \|f\|_1$. To define $\hat{f}$ for $f \in L^2$, we need to use trickery and deceit.

3

The space of rapidly decreasing or Schwartz functions is denoted by S and is defined as the class of all smooth functions f on $\mathbb{R}^n$ for which the seminorms $\sup_x \left| x^\alpha D^\beta \phi(x) \right|$ are finite. We've used multiindex notation here; if we let α denote the multiindex $\alpha = (\alpha_1, \alpha_2, \dots, \alpha_n)$; then $x^\alpha = x_1^{\alpha_1} \cdots x_n^{\alpha_n}$. As these seminorms vary with α, they give rise to a topology on S. The dual of S is then called the space of tempered distributions. Then it is not hard to prove:

0.2 THEOREM.
 a) S is dense in L^p if $1 \le p < \infty$.
 b) The Fourier transform is a continuous, one to one map of S onto S.
 c) If the convolution of two functions f and g in S is defined as

$$f \star g = \int f(x-y)g(y)dy,$$

then this function is again in S and $\widehat{f \star g} = \hat{f}\hat{g}$.
 d) If f and g are in S, then $\int f\bar{g} = \int \hat{f}\bar{\hat{g}}$
 e)
$$\|f\|_2 = \|\hat{f}\|_2$$

 f) The Fourier transform of a distribution u is defined by $\hat{u}(f) = u(\hat{f})$. Then $\hat{u}$ is again a distribution.
 g) The translation operator τ_y is defined by $(\tau_y f)(x) = f(x-y)$. If convolution of a distribution u and a function in S is defined as $(u \star f)(x) = u(\tau_x \tilde{f})$, where $\tilde{f}(y) = f(-y)$, then $u \star f$ is a smooth function.

The space S is a technical tool which makes it easy to do computations with integrals that might otherwise be infinite. We will also need some notation for the rest of the book. If T is a bounded linear operator from L^p to L^p, we denote its operator norm by $_p\|T\|$. Recall then that

$$_p\|T\| = \sup\{\|Tf\|_p \mid \|f\|_p = 1\}$$

and that

$$_p\|T\| = \sup\{\int Tfg \mid \|f\|_p = \|g\|_{p'} = 1\},$$

where $\frac{1}{p} + \frac{1}{p'} = 1$.

0.4 NOTES FOR CHAPTER 0

0.1): Bochner-Riesz means are a variant of Riesz means, $(1 - |n|^2)_+^\alpha$; we shall see that the two summability methods give the same results in L^p, but that Bochner-Riesz means are much easier to compute with. The original interest of the S_R^α was in conection with pointwise summability and the localization of Fourier series (a topic now in disrepute due to its difficulty). Bochner's paper [1] showed that n-dimensional problems are deeper than one-dimensional; localization did not hold for all α, in contrast to one-dimensional results. The essential ideas are also treated in Stein and Weiss [50] Chapter 7 Theorem 4.2.

There are good reasons for grouping Fourier coefficients together in concentric spheres rather than in concentric polygons. In Bochner's words([**1**] pps. 179-80):
"The elementary exponentials $u(x) = e^{i(n_1 x_1 + \cdots + n_k x_k)}$ (all $n_1, \ldots, n_k$ integers) are a complete set of regular solutions of the characteristic value problem

$$\Delta u(x) = -\lambda u(x),$$

if this system is being considered on the (closed) torus

$$0 \leq x_1 \leq 2\pi, \ldots, 0 \leq x_k \leq 2\pi,$$

and Δ is the Laplace operator with respect to the Euclidean metric on the torus, namely

$$\frac{\partial^2 u}{\partial x_1^2} + \cdots + \frac{\partial^2 u}{\partial x_k^2}.$$

Since $\lambda = n_1^2 + \cdots + n_k^2$ our way of writing series . . . . satisfies the very natural principal of ordering the terms in a series according to the magnitude of the characteristic eigenvalues λ."

To paraphrase Bochner, in looking at spherical summability methods, we are really looking at the question of the convergence of the eigenfunction expansion of a differential operator on a compact Riemannian manifold. C. Fefferman's result has been used to show that if $n > 1$, such expansions never converge. This is our main point: the divergence of spherical means are not an abberation caused by a crazy choice of summability methods, but is a fundamental fact of high dimension geometry.

The results summarized in Figures 1 and 2 are due to many individuals. See Herz [**27**], C. Fefferman [**23**], Carleson and Sjolin [**5**], and Tomas [**51**] for a summary.

0.2): The computation of the $\|K_R\|_1$, which are called the Lebesgue constants, is due to Fejer (1910). These sort of techniques fail miserably in higher dimensions, because the series for $K_R(\theta)$ cannot be summed. It is possible to show that $\|K_R\|_1 = 0(R^{\frac{n-1}{2}})$ if $n \geq 2$, which is the best possible estimate; see Shapiro [**44**] for the proof and references to earlier results. In trying to get an explicit formula for the partial sums K_R, you quickly realize that a really good formula would allow you to compute exactly the number of integer points inside a ball of radius R in $\mathbb{R}^n$. Unfortunately, getting good estimates on this is one of the hardest problems in number theory. It is a serious problem because for some R, there are no pairs of integers (m, n) with $m^2 + n^2 = R^2$. But for a slightly different R, there will be lots of pairs. You can't expect any really regular expression that would follow from a good formula for K_R.

The classical, detailed proofs of the divergence of Fourier series at a point are best found in Zygmund [**57**], Chapter 7 Volume 1.

CHAPTER 1 MULTIPLIER THEORY

SECTION 1.1 MULTIPLIERS ON L^p

There are problems "summing" Fourier integrals; the purpose of this section is to see what the problems are and how they can be overcome.

If $f \in L^2$, the expected inverse Fourier transform

$$\check{f} = \int f(\xi) e^{2\pi i x \xi} d\xi$$

simply does not converge. The traditional remedy is to look instead at "partial sums" : integrals over bounded regions; this forces the integral to converge. One then hopes to take a limit and have convergence of the limit. We define the partial sums operators S_R by

$$S_R f(x) = \int_{|\xi| \leq R} \hat{f}(\xi) e^{2\pi i x \xi} d\xi,$$

which does converge absolutely. We hope that for all $f \in L^2$, $\lim_{R \to \infty} S_R f$ exists in L^2, and that the limit is f. The first problem is that we don't even know that $S_R f$ is in L^2, let alone convergent in L^2. So we begin with the functional analysis of the operators S_R: do they take L^2 functions into L^2 functions? We let B denote the unit ball in $\mathbb{R}^n$, and let $\chi_B \left(\frac{\xi}{R} \right)$ be the characteristic function of the ball of radius R. It follows that

$$S_R f(x) = \int \chi_B \left(\frac{\xi}{R} \right) \hat{f}(\xi) e^{2\pi i x \xi} d\xi.$$

These are the operators we intend to study.

1.1 DEFINITION: A Fourier multiplier operator on $L^p(\mathbb{R}^n)$ is a linear operator T bounded on L^p for which there exists a $\mu \in L^\infty(\mathbb{R}^n)$ satisfying

$$T f(x) = \int \mu(\xi) \hat{f}(\xi) e^{2\pi i x \xi} d\xi$$

for all Schwartz functions f. In this case, μ is called the Fourier multiplier associated to the operator T.

REMARK: The operator is really defined by a functional equation: $\widehat{Tf} = \mu \hat{f}$, which is very indirect. To actually work with multiplier operators, we need characterizations which stay on one side: either the function side or the Fourier transform side.

1.2 PROPOSITION. *Assume T is a Fourier multiplier of L^p. Then there is a distribution K for which*

$$T f(x) = K \star f(x),$$

for all $f \in \mathcal{S}$. K is called the convolution kernel of T.

PROOF: To see what is going on, take $\mu \in \mathcal{S}$. Then

$$Tf(x) = \int \mu(\xi)e^{2\pi i x\xi}\hat{f}(\xi)d\xi = \int \mu(\xi)e^{2\pi i x\xi}\int e^{-2\pi i\xi y}f(y)dy$$

$$= \int\int \mu(\xi)e^{2\pi i\xi(x-y)}d\xi f(y)dy = \int \check{\mu}(x-y)f(y)\ dy = \check{\mu} \star f(x).$$

Because everything in sight was a Schwartz function, we could rearrange the integrals, and we took the distribution K to be the function $\check{\mu}$. To give a general proof, we need a lot of distribution facts. First of all, if $f \in \mathcal{S}$, $\hat{f} \in \mathcal{S}$ and, since $\mu \in L^\infty$, $\mu\hat{f} \in L^1$, so that

$$Tf(x) = \int e^{2\pi i x\xi}\mu(\xi)\hat{f}(\xi)\ d\xi$$

exists for all x.

The distribution K is defined by $K(f) = (T\tilde{f})(0)$. The integral representation for f shows that this is a distribution: $f \to \hat{f}$ is a continuous map of $\mathcal{S}$ into $\mathcal{S}$, and integration against an L^∞ function is well known to be a distribution. To finish, notice that

$$(K \star f)(x) \equiv K(\tau_x\tilde{f}) \equiv T(\widetilde{\tau_x\tilde{f}})(0)$$

$$= T(\tau_{-x}f)(0) = \int \mu(\xi)\widehat{\tau_{-x}f}(\xi)d\xi = \int \mu(\xi)e^{2\pi i x\xi}\hat{f}(\xi)d\xi \equiv Tf(x);$$

the next to the last equality holds because

$$\widehat{\tau_x f}(\xi) = \int e^{-2\pi i y\xi}(\tau_x f)(y)dy = \int e^{2\pi i x\xi}\hat{f}(y).$$

1.3 REMARKS:

a) As distributions, $\hat{K}$ is μ and $\check{\mu}$ is K.

b) The distribution K could be very bad; look at the operator $Tf = f$. Here the Fourier inversion formula tells me that $\mu \equiv 1$. What distribution gives this? The distribution corresponding to T is the Dirac delta measure, $\delta(f) = f(0)$. Working out the formalities,

$$(\delta \star f)(x) \equiv \delta(\tau_x\tilde{f}) = (\tau_x\tilde{f})(0) = \tilde{f}(-x) = f(x).$$

The moral is that even trivial operators can generate distributions that are not even given by integration against functions. To make things worse, we will now sketch an example where K is not even given by integration against a finite measure. Let μ be the L^∞ function $i\,sign(\xi)$, then we will compute in 2.13 that $\check{\mu}(x) = \frac{1}{\pi x}$. But let's face it, $\check{\mu}(f) = \frac{1}{\pi}\int \frac{1}{x}f(x)dx$ normally does not converge. Properly speaking, we have to take a principal value integral integral,

$$\lim_{\epsilon \to 0} \frac{1}{\pi}\int_{|x|\geq\epsilon} \frac{1}{x}f(x)dx.$$

This sort of distribution should be expected when the multiplier is not an integrable function.

8

1.4 PROPOSITION. *If T is a multiplier of L^p and $f \in \mathcal{S}$, then*

$$(\widehat{Tf})(\xi) = \mu(\xi)\hat{f}(\xi)$$

PROOF: To see this, we have to see that as distributions, $\widehat{Tf}$ and $\mu(\xi)\hat{f}(\xi)$ act the same. Pointwise equality follows immediately, since both distributions are given by integration against functions; that was the whole point of taking $f \in \mathcal{S}$; then $\hat{f} \in \mathcal{S}$ and $\mu\hat{f} \in L^1$. It follows that Tf is even continuous. Ok, checking equality by integrating against a function $g \in \mathcal{S}$,

$$\int \widehat{Tf}g = \int Tf\hat{g},$$

from 0.2 extended to all of L^2. But

$$\int \mu\hat{f}g = \int\int \mu\hat{f}(g) = \int (\mu\hat{f})\hat{g} = \int Tf\hat{g}.$$

The triple (T, K, μ), is called a multiplier triple. T is the multiplier operator, K the convolution kernel, and μ the multiplier. In the rest of the book we will often use phrases like " Let T be an operator with multiplier μ", or, " If μ is a multiplier with kernel K".

1.5 PROPOSITION. *Let T be a multiplier operator on L^2, with multiplier μ. Then*

$$_2\|T\| = \|\mu\|_\infty.$$

PROOF: From the Plancherel theorem,

$$\|Tf\|_2 = \|\widehat{Tf}\|_2 = \|\mu\hat{f}\|_2$$

$$\leq \|\mu\|_\infty\|\hat{f}\|_2 = \|\mu\|_\infty\|f\|_2 ,$$

so that $_2\|T\| \leq \|\mu\|_\infty$. Now we'll show that if $\epsilon > 0$, then $_2\|T\| \geq \|\mu\|_\infty - \epsilon$.

From the definition of L^∞, the set $E = \{\xi \mid |\mu(\xi)| \geq \|\mu\|_\infty - \epsilon\}$ has positive measure; we can choose a subset S of E with positive but finite measure. Then we can compute the norm of this function: $\chi_S \in L^2$, and $\|\chi_S\|_2 = \|\hat{\chi}_S\|_2 = |S|^{\frac{1}{2}}$. We also can compute the norm of T applied to this function:

$$_2\|T\| \, \|\check{\chi}_S\|_2 = \,_2\|T\| \, |S|^{\frac{1}{2}} \geq \|T(\hat{\chi}_S)\|_2 = \|\mu\chi_S\|_2$$

$$\geq \inf |\mu(\xi)\chi_S(\xi)|\|\chi_S\|_2 \geq (\|\mu\|_\infty - \epsilon)|S|^{\frac{1}{2}} = (\|\mu\|_\infty - \epsilon)\|\check{\chi}_S\|_2.$$

1.6 TECHNICAL REMARK: We used the result that $\widehat{T(\hat{\chi}_S)} = \mu\chi_S$. We really only proved this if $f \in \mathcal{S}$, but the characteristic function of a set S is certainly not in $\mathcal{S}$. This is typical of the petty technical problems that plague this subject, all due to the fact that our multipliers are only defined on a dense set of L^p, and extended to

9

the remainder by continuity. In this remark we want to show how to treat a typical petty problem.

For $f \in L^2$, choose $f_j \in S$ such that f_j converges to f in L^2. Since the Fourier transform is continuous on L^2, $\hat{f}_j$ converges to $\hat{f}$ in L^2. Since T is a bounded(that is, continuous) operator on L^2, Tf_j converges in L^2 to Tf, and $\widehat{Tf_j}$ converges to $\widehat{Tf}$ in L^2. We may take a subsequence j_k such that $\widehat{Tf_{j_k}}$ converges and $\hat{f}_{j_k}$ converges pointwise ae to $\widehat{Tf}$ and to $\hat{f}$. Then, for almost all ξ,

$$\widehat{Tf}(\xi) = \lim_{k \to \infty} \widehat{Tf_{j_k}}(\xi) = \lim_{k \to \infty} \mu(\xi)\hat{f}_{j_k}(\xi) = \mu(\xi)\hat{f}(\xi).$$

1.7 PROPOSITION. *T is a multiplier of L^1 if and only if the convolution kernel K is a finite measure, dm, and in that case*

$$_1\|T\| = \|dm\|.$$

PROOF: This result has a simple intuition. The Dirac delta function δ is almost an L^1 function, and it also acts like the identity operator. Then $_1\|T\|$ should be like $\|K \star \delta\|_1 = \|K\|_1$ This proof shows how to handle the "minor" technical annoyance that δ is not a function in L^1. If we assume that K acts as a finite measure, then

$$\|K \star f\|_1 = \|K(\tau_x \tilde{f})\|_1 = \left\| \int (\tau_x \tilde{f})(y) dm(y) \right\|_1$$

$$\leq \int \|\tau_x \tilde{f}\|_{1,y} d|m|(y) = \int \|f(y-x)\|_1 d|m|(y) = \|f\|_1 \|dm\|.$$

It follows that $_1\|T\| \leq \|dm\|$.

For the other direction, I need to get at the δ function even though it is not an L^1 function. We'll use an absolutely standard technique: instead of using δ, we approximate it by functions which are in L^1. We choose Gaussian kernels:

$$\omega_t(x) = (2\pi t)^{-n} \exp\left[-(|x|/2t)^2 \right];$$

we rigged it so that $\|\omega_t\|_1 = 1$, $\omega_t \in S$, and so that for every $f \in S, \omega_t \star f$ converges in S to f as t tends to zero. Of course we get

$$\|T(\omega_t)\|_1 \leq \,_1\|T\|\|\omega_t\|_1 = \,_1\|T\|;$$

the question is whether we get $T(\omega_t) = K \star \omega_t$ converging to dm. If K were a function in S, this would be no problem; here all we know is that the $T(\omega_t)$ are uniformly bounded in L^1. To get from this sort of information(boundedness) to convergence information is clearly some sort of compactness condition. Unfortunately, the unit ball in L^1 is not sequentially compact, so we resort to trickery and deceit.

The $T(\omega_t)$ are in L^1, and they can be viewed as finite measures with total variation norm bounded by 1. Amazingly enough there is a topology on the space of finite measures, $\mathcal{M}$ under which closed, bounded sets are sequentially compact. We view $\mathcal{M}$ as the dual of $\mathcal{C}_0$, the space of countinuous functions vanishing at infinity. The

10

topology we put on $\mathcal{M}$ is called the weak-$\star$ topology: μ_j converges to μ means that $\mu_j(f)$ converges to $\mu(f)$ for all $f \in \mathcal{C}_0$. Then Alogolu's theorem (Rudin [**43**]) asserts the needed compactness: there is a subsequence $T(\omega_{t_j})$ which converges weak-$\star$ to some finite measure dm in the ball of radius $_1\|T\|$. We've gotten as far as knowing what dm is, and we even know that the total variation norm of dm is bounded by $_1\|T\|$. The only little detail lacking in this sweet picture is the statement that the distribution K is given by integration against dm. This is the role of the weak-$\star$ convergence. Since $\mathcal{S} \subset \mathcal{C}_0$, for every $f \in \mathcal{S}$,

$$dm(f) \equiv \int f \, dm = \lim_{j \to \infty} \int f T(\omega_{t_j}) = \lim_{j \to \infty} \int f K \star \omega_{t_j}$$

$$= \lim_{j \to \infty} \int f(x) K \left(\tau_x \tilde{\omega}_{t_j}(y) \right) dx = \lim_{j \to \infty} K \left(\int f \omega_{t_j}(y - x) dx \right).$$

The last equality is true because the integral can be approximated by a Riemann sum. The approximation will converge in $\mathcal{S}$ and so we interchanged the approximation and the distribution. Finally,

$$\lim_{j \to \infty} K \left(\omega_{t_j} \right) \star f = K \left(\lim_{j \to \infty} \omega_{t_j} \star f \right) = K(f).$$

This is called a "regularity" result; at first we only knew that K was some potentialy awful distribution. In fact, it had to be a much nicer object, a finite measure. How did a distribution suddenly get forced to be a measure? The boundedness of an L^1 norm forced a sequence of smooth approximations to converge. So this is our first example of how L^p boundedness of operators forces regularity; once again note the role of functional analysis.

1.8 LEMMA. *If T is a multiplier of L^p, then it is a multiplier of $L^{p'}$, where $\frac{1}{p} + \frac{1}{p'} = 1$.*

PROOF: To see the idea here, remember that $L^{p'}$ is the dual space of L^p, so that the boundednes of the operator T on L^p implies boundedness of the adjoint operator T^* on $L^{p'}$. Our goal is to relate T and T^*. If T were given by convolution with some sort of reasonable function, say a nice Schwartz function K, then we could compute:

$$\int Tf\bar{g} = \int K \star f\bar{g} = \int \int K(x-y)f(y)\bar{g}(x)dydx$$

$$= \int \int K(x-y)\bar{g}(x)f(y)dydx = \int \left[K \star (\widetilde{\bar{g}}) \right] f(y)dy$$

$$= \int f(y)\widetilde{K \star (\widetilde{\bar{g}})}(y)dy.$$

This computation means that if we define the isometry J of L^p by $Jf = \bar{\tilde{f}}$, then $T^* = JTJ$. Since J is an isometry, we get

$$_p\|T\| = {}_{p'}\|T^*\| = {}_{p'}\|T\|.$$

The real problem will be in changing the order of integration when K is not a nice Schwartz function. We will do some unpleasant computations with distributions, following the intuition we just gave.

$$\int Tf\bar{g} = \int K \star f(x)\bar{g}(x)dx = \int K\left(\tau_x(\tilde{f})(y) \right) \bar{g}(x)dx$$

$$= K\left(\int (\tau_x\tilde{f})(y)\bar{g}(x)dx \right) = K\left(\int f(x-y)\bar{g}(x)dx \right)$$

$$= K\left(\int f(x)\bar{g}(x+y)dx \right) = K\left(\int f(x)(\tau_{-x}\bar{g})(y)dx \right)$$

$$= \int f(x) \left[K \star \tilde{\bar{g}}(x) \right] dx.$$

Etc, etc.

1.10 LEMMA. *If $K \in L^1$, then $\|K \star f\|_p \leq \|K\|_1 \|f\|_p$.*

PROOF: As in the proof of 1.7,

$$\|K \star f\|_p = \| \int K(y)f(x-y)dy\|_p$$

$$\leq \int |K(y)| \, \|f(x-y)\|_{p,x}dy = \|K\|_1 \|f\|_p.$$

1.11 REMARK: We now have a pretty complete understanding of multipliers on L^1 and L^2; we would like to get a better understanding of L^p for $1 < p < 2$. By duality, that is, by 1.8, we would get a grasp of multipliers of L^p for $2 < p < \infty$.

Think about the identity operator, bounded on all L^p. If we understand a function in L^1 and in L^2, why should that tell us anything about the function in L^p for $1 < p < 2$? What is needed here is some sort of convexity: if you think of things graphically, the operator norm of T on L^p should lie below the norm for T on the line between L^1 and L^2. This happens not to be true. Look at Holder's inequality:

$$\int |f|^{tp_0+(1-t)p_1} \leq \left(\int |f|^{p_0} \right)^t \left(\int |f|^{p_1} \right)^{1-t} ;$$

this tells me that $log \left(\int |f|^p \right)$ is a convex function of p.

1.12 THEOREM. *Let $S = \{z \in \mathbb{C} \mid 0 \leq Re(z) \leq 1\}$. Let F be bounded and continuous on S, and analytic in the interior of S. Let $k_x = \sup_y |F(x+iy)|$. Then $\log k_x$ is a convex function of x.*

IDEA OF PROOF: This is a standard result from complex variable theory; the first thing to do is to rescale F to be bounded by 1 on $0+iy$ and on $1+iy$; the rescaled function is $G(z) = F(z)k_0^{-z}k_1^{1-z}$. Then we try to prove $|G| \leq 1$ everywhere. If G vanishes as $|y| \to \infty$, we rig up a rectangle on whose boundary $|G| \leq 1$, outside of which $|G| \leq 1$. Inside, we apply the maximum principle. Now if G does not vanish at infinity, we use the boundedness of G to multiply by an exponential tending to zero. As we let the exponential go to infinity more slowly, we recover G. It follows that $G(z)$ is everywhere bounded by 1, and therefore that $|F(z)| \leq k_0^z k_1^{z-1}$. Taking the supremum and then the log gives the right result.

1.13 THEOREM. *Let T be a linear operator bounded on L^{p_i}, $i = 0, 1$. Then $\log {}_p\|T\|$ is a convex function on $\left[\frac{1}{p_0}, \frac{1}{p_1} \right]$. That is, if $\frac{1}{p} = \frac{t}{p_0} + \frac{1-t}{p_1}$ for some t with $0 \leq t \leq 1$, then*

$${}_p\|T\| \leq {}_{p_0}\|T\|^t \, {}_{p_1}\|T\|^{1-t} .$$

PROOF: Clearly we want to use 1.12, and take $F(z) = \int |Tf|^z$. This runs into technical problems at any x for which $Tf(x) = 0$, and I can't raise $T(f)$ above 0 by adding on ϵ without destroying finiteness of the integrals. Then too, this doesn't get me a look at the ${}_p\|T\|$. I fix all these problems in one step: I'll test the size of $T(f)$ by integrating it against an $L^{p'}$ function, and I'll avoid the zeroes problem by evaluating T on simple functions. Thus, let f, g be simple functions with

$$\|f\|_p = \|g\|_{p'} = 1,$$

$$f(x) = \sum |a_j|e^{i\theta_j}\chi_{E_j}; \quad g(x) = \sum |b_k|e^{i\phi_k}\chi_{F_k}.$$

Here E_j, F_k are sets of finite measure; we will show that

$$\int Tfg \leq {}_{p_0}\|T\|^t \, {}_{p_1}\|T\|^{1-t}.$$

First let $\alpha(z) = \frac{z}{p_0} + \frac{1-z}{p_1}$. This maps the $\frac{1}{p}$ interval into $[0, 1]$. Then let

$$F(z) = \int T \left(\sum |a_j|^{p\alpha} e^{i\theta_j} \chi_{E_j} \right) \left(\sum |b_k|^{p'(1-\alpha)} e^{i\phi_k} \chi_{F_k} \right).$$

13

This is not the same as raising Tf to the z power, but it is an analytic function of z in the strip S, and bounded and continuous. Note that

$$|F(x+iy)| \leq \ _{p_0}\|T\| \ \| \sum |a_j|^{p\alpha} e^{i\theta_j} \chi_{E_j} \|_{p_0} \ \| \sum |b_k|^{p'(1-\alpha)} e^{i\phi_k} \chi_{F_k} \|_{p'}$$

But

$$\| \sum |a_j|^{p\alpha} e^{i\theta_j} \chi_{E_j} \|_{p_0} = \| \sum |a_j|^{Re(p\alpha)} \chi_{E_j} \|_{p_0}$$

because of the disjointness of the sets E_j. Now

$$Re(\alpha p) = pRe \left[\frac{1-iy}{p_0} + \frac{iy}{p_0} \right];$$

it follows that

$$\| \sum |a_j|^{Re(p\alpha)} \chi_{E_j} \|_{p_0} = \| \sum |a_j|^{\frac{p}{p_0}} \chi_{E_j} \|_{p_0}$$

$$= \| \sum |a_j| \chi_{E_j} \|_p^{\frac{p}{p_0}} = \|f\|_p^{\frac{p}{p_0}} = 1.$$

Similar estimates apply to the second term, and it follows that

$$|F(0+iy)| \leq \ _{p_0}\|T\|; \ |F(1+iy)| \leq \ _{p_1}\|T\|.$$

The Theorem now follows from 1.12.

REMARK: This result is called the Riesz-Thorin interpolation theorem.

1.14 THEOREM. If $1 \leq p \leq 2$, $\|\hat{f}\|_{p'} \leq \|f\|_p$.

PROOF:

$$\|Tf\|_\infty = \|\hat{f}\|_\infty \leq \|f\|_1$$

$$\|Tf\|_2 = \|\hat{f}\|_2 = \|f\|_2$$

These two cases show that T maps L^p into $L^{p'}$ for the special cases of $p = 1$ and $p = 2$. The p, p' relationship is linear; interpolation preserves it. The map has norm bounded by 1 on each of the endpoint spaces, and interpolation preserves this norm as well.

1.15 REMARKS:

a) Initially, the Fourier transform was defined only for L^1 functions, because only there did the integrals converge absolutely. The Plancherel theorem allowed us to extend the operator to L^2, but if $f \in L^2$, $\hat{f}$ is only defined almost everywhere. The result above is called the Hausdorff-Young theorem, and it allows us to extend the definition to L^p for $1 \leq p \leq 2$. Thus, the Fourier transform of a function in L^p is the limit in $L^{p'}$ of the Fourier transform of Schwartz functions.

b) The Hausdorff-Young inequality is pretty much the best we can do, at least if size is measured by L^p norms. The following trick demonstrates it: let $f_\delta(x) = f(\delta x)$. Then for a fixed f,

$$\|f_\delta\|_p = \|f\|_p \delta^{-\frac{n}{p}} = c\delta^{-\frac{n}{p}}$$

$$\hat{f}_\delta(\xi) = \delta^{-n} \hat{f}(\delta^{-1}\xi)$$

$$\|\hat{f}_\delta\|_q = c'\delta^{-\frac{n}{q'}}.$$

Therefore, $\|\hat{f}\|_q \leq C\|f\|_p$ can only happen for some f if

$$c'\delta^{-\frac{n}{q'}} \leq c\delta^{-\frac{n}{p}}$$

for all $\delta > 0$. Thus $q' = p$.

14

1.16 LEMMA. *Let (T, K, μ) be a multiplier of L^p. Then*

$$\|\mu\|_\infty \leq {}_p\|T\|.$$

PROOF: Note that ${}_p\|T\| = {}_{p'}\|T\|$, so that we can interpolate between L^p and $L^{p'}$. L^2 always lies in between, so the operator is always bounded on L^2. Moreover, $\frac{1}{2} = \frac{1}{2}\left(\frac{1}{p} + \frac{1}{p'}\right)$, whence

$$\|\mu\|_\infty = {}_2\|T\| \leq {}_p\|T\|^{\frac{1}{2}} \, {}_{p'}\|T\|^{\frac{1}{2}} = {}_p\|T\|.$$

1.17 REMARK: Notice that ${}_p\|T\| \leq \|K\|_1$, as remarked in 1.10, although for a multiplier of L^p the L^1 norm of the convolution kernel is probably not finite. But we now do have a relationship amongst all three components of a multiplier triple, (T, K, μ):

$$\|\mu\|_\infty \leq {}_p\|T\| \leq \|K\|_1.$$

Of course, this relation is worthless for most purposes: we can expect the estimate to be good only when the extreme sides are equal: $\|\mu\|_\infty = \|K\|_1$. But this means $\|\hat{K}\|_\infty = \|K\|_1$. But $\|\hat{K}\|_\infty \leq \|K\|_1$ in general. On the other hand, if K is a positive function,

$$\|K\|_1 = \int K = \int K(x)e^{2\pi i 0 x}dx = \hat{K}(0) \leq \|\hat{K}\|_\infty \ ;$$

so that for positive K,

$$\|\hat{K}\|_\infty = \|K\|_1 = {}_p\|T\|.$$

For positive convolution kernels, then, the whole story of multipliers is clear: they yield multipliers of an L^p if and only if they yield multipliers of all L^p, and in this case their L^1 norm is exactly their L^p operator norm. On the other hand, we saw in the Introduction that really the most interesting multiplier operators have terribly unpositive convolution kernels. For such multipliers, the gap between $\|K\|_1$ and $\|\hat{K}\|_\infty$ is large, but we will need to control both of these to control the operator norm of T.

This is one of the central problems in Fourier analysis: the need for simultaneous control of K and $\hat{K}$.

1.18 REMARK: There are other means of controlling functions in L^p for $1 < p < 2$. The simplest is to chop an L^p function into pieces, each of which is in some other Lebesgue class. The basic example of this is setting

$$G = \{x| \ |f(x)| \leq 1\}; \ B = \{x| \ |f(x)| > 1\}.$$

Then $f = f\chi_G + f\chi_B = g + b$. This is a decomposition of f into a "good" function g and a "bad" function b; the good part is that

$$\|g\|_2^2 = \int_G |f|^2 = \int_G |f|^{2-p}|f|^p \leq \int_G |f|^p \leq \|f\|_p^p;$$

$$\|b\|_1 = \int |b| = \int_B |f| \leq |B|^{\frac{1}{p}}\|f\|_p.$$

15

So we see that an arbitrary function in L^p for p between 1 and 2 can be written as a sum of functions, one in L^1 and the other in L^2. If T is a linear operator, we have some hope of following the L^p boundedness of T simply knowing the L^1 and L^2 behaviour. Now we'll do all this more carefully.

1.19 DEFINITION: A measurable function is in weak L^p if

$$|\{x \mid |f(x)| > \lambda\}| \leq C\lambda^{-p}$$

where C is independent of λ, $0 < \lambda < \infty$.

An operator T is said to be weak (p, p) if

$$|\{x \mid |Tf(x)| > \lambda\}| \leq C\frac{\|f\|_p^p}{\lambda^p}$$

where C is independent of λ and f.

1.20 REMARKS:

a) Usually we write $|\{f > \lambda\}|$ as a shorthand for $|\{x \mid |f(x)| > \lambda\}|$; since we may often take positive functions, it turns out this never causes notational problems. This is called the distribution function of f, and often written as $\lambda(f)$.

b) A typical function which is in weak L^1 but not in L^1 is $f(x) = \frac{1}{x}$. A standard computation shows L^p functions are always weak L^p:

$$\lambda^p|\{x \mid |f(x)| > \lambda\}| = \int_{\{f>\lambda\}} \lambda^p \, dx$$

$$\leq \int_{\{f>\lambda\}} |f(x)|^p \, dx \leq \int |f|^p = \|f\|_p^p.$$

c) If weak L^p is going to be any good in analyzing serious L^p functions, distribution functions need to be connected with L^p norms. The relation is:

$$\|f\|_p^p = p \int_0^\infty |\{x \mid |f(x)| > \lambda\}| \, \lambda^{p-1} d\lambda.$$

The simplest proof uses simple functions, and this reduces the whole thing to what happens for $f = \chi_I$, where I is an interval. Then $\|f\|_p^p = |I|$, and $|\{x \mid |f(x)| > \lambda\}| = 0$ if $\lambda > 1$, and equals $|I|$ if $0 < \lambda \leq 1$. Then the integral is

$$p|I| \int_0^1 \lambda^{p-1} d\lambda = |I|.$$

1.21 THEOREM. Assume T is a linear operator which is weak (p_0, p_0) and is weak (p_1, p_1). Then T is a bounded operator on L^p for p between p_0 and p_1.

PROOF: To make it easy on the authors, we do only the case $1 \leq p_0 < p < p_1 < \infty$. As in Remark 1.18, we let:

$$G = \{|f| \leq \frac{\lambda}{2}\}, \quad B = \{|f| > \frac{\lambda}{2}\}.$$

16

$$g = f\chi_G; \quad b = f\chi_B$$

and we note that $|Tf| = |Tg + Tb| \le |Tg| + |Tb|$. Unfortunately, distribution functions are not linear, so this decomposition of Tf is not immediately useful. But there is a substitute for linearity:

$$|\{|Tf| > \lambda\}| \le |\{|Tg| + |Tg| > \lambda\}|$$

$$\le |\{|Tg| > \frac{\lambda}{2}\}| + |\{|Tb| > \frac{\lambda}{2}\}|$$

$$\le C_0 \lambda^{-p_0} \int |b|^{p_0} + C_1 \lambda^{-p_1} \int |g|^{p_1}.$$

Now we compute:

$$\|Tf\|_p^p = p \int |\{|Tf| > \lambda\}| \lambda^{p-1} d\lambda$$

and find that it is bounded by the sum of two terms; the first term is:

$$C_0 p \int_0^\infty \lambda^{p-p_0-1} \int_{\{|f| \ge \frac{\lambda}{2}\}} |f|^{p_0} dx d\lambda$$

$$= C_0 p \int \int_0^{2|f|} \lambda^{p-p_0-1} d\lambda |f(x)|^{p_0} dx = \frac{2^{p-1} p C_0}{p - p_0} \|f\|_p^p.$$

Similarly, the second term is bounded by:

$$\frac{2^{p-1} p C_1}{p_1 - p} \|f\|_p^p.$$

REMARK: This result is due to Marcinkiewicz; notice that as p tends to p_0 or to p_1, the bounds on the multiplier get very bad. The rate at which they get bad provides extra information at the endpoints. The basic example to keep in mind is $\frac{1}{x}$; here the L^1 norm is not finite, but the divergence is measured by $\log x$. A similar result holds in the Marcinkiewicz interpolation theorem; see the notes at the end of the Chapter.

1.22 LEMMA. *If f, g are in S, then*
  a) $\widehat{\bar{f}}(\xi) = \overline{\hat{f}(-\xi)}$.
  b) *If $f_\delta(x) = f(\delta x)$, then $\hat{f}_\delta(\xi) = \delta^{-n} \hat{f}(\delta^{-1}\xi)$.*
  c) *If R denotes a rotation of $\mathbb{R}^n$, and $f_R(x) = f(Rx)$, then $\hat{f}_R(\xi) = \hat{f}(R^{-1}\xi)$.*
  d) $(\widehat{\tau_y f})(\xi) = e^{2\pi i y \xi} \hat{f}(\xi)$.
  e) $\widehat{(fg)}(\xi) = \hat{f} \star \hat{g}(\xi)$.
  f) $(f \star g)\hat{}(\xi) = \hat{f}(\xi)\hat{g}(\xi)$.
  g) $D_{\xi_j}\hat{f}(\xi) = -2\pi i (x_j f)\hat{}(\xi)$.

PROOF: All of these follow from changes of variables and the definitions. A reader who has not seen these results before is urged to work out the details. This is making friends with the Fourier transform.

17

1.23 THEOREM. *Assume* (T, μ, K) *is a Fourier multiplier triple for* $L^p(\mathbb{R}^n)$. *Then:*

a) $Re(\mu)$ *and* $Im(\mu)$ *are* L^p *multipliers.*

b) Let T_δ *be the operator with multiplier* $\mu(\delta\xi)$. *Then* $_p\|T_\delta\| = {}_p\|T\|$.

c) Let T_η *be the operator with multiplier* $\mu(\xi - \eta)$. *Then* $_p\|T_\eta\| = {}_p\|T\|$.

d) If $\check{\psi} \in L^1$, *let* T_ψ *be the operator with multiplier* $\psi\mu$. *Then* $_p\|T_\psi\| \le \|\check{\psi}\|_1 {}_p\|T\|$.

e) If $\psi \in L^1$, *let* T_ψ *be the operator with multiplier* $\psi \star \mu$. *Then* $_p\|T_\psi\| \le \|\psi\|_1 {}_p\|T\|$.

f) Let $\xi_1, \xi_2, \ldots, \xi_{n-k}$ *be fixed, and define*

$$\mathbb{R}^{k_0} = \{\xi \in \mathbb{R}^n | \xi_1 \cdot \xi + \ldots + \xi_{n-k} \cdot \xi = \lambda\}.$$

Let $\mu_0 = \mu|_{\mathbb{R}^{k_0}}$ *and let* T_0 *be the operator on* $L^p(\mathbb{R}^{k_0})$ *with multiplier* μ_0. *Then for almost all* λ, $_p\|T_0\| \le {}_p\|T\|$.

PROOF: The proofs of a) - d) are all easy, and all make use of the fact that the Fourier transform behaves predictably under conjugation, dilation, rotation, translation, or what-have-you. We do a typical proof to suggest that the reader have the same experiences:

d)

$$T_\eta f(x) = \int \mu(\xi - \eta)e^{2\pi i x\xi}\hat{f}(\xi)d\xi$$

$$= \int \mu(\xi)e^{2\pi i x(\xi+\eta)}\hat{f}(\xi + \eta)d\xi$$

$$= e^{2\pi i x\xi\eta}\int \mu(\xi)e^{2\pi i x\eta}\left(e^{-2\pi i x\eta}f(x)\right)(\xi)d\xi = JTJ^{-1}(f)$$

where J is the isometry of multiplication by $e^{2\pi i x\eta}$.

e) We will use duality, a very standard trick in the Fourier multiplier game. Choose f, g, in $\mathcal{S}$ with $\|f\|_p = \|g\|_{p'} = 1$, and compute:

$$\int T_\psi f\bar{g} = \int \widehat{T_\psi f}\,\hat{\bar{g}} = \int \psi \star \mu\hat{f}\,\hat{\bar{g}}$$

$$= \int\int \psi(\eta)\mu(\xi - \eta)\hat{f}(\xi)\,\hat{\bar{g}}(\xi)d\eta d\xi = \int \psi(\eta)\int \mu(\xi - \eta)\hat{f}(\xi)\,\hat{\bar{g}}(\xi)d\xi d\eta$$

$$\le \|\psi\|_1 \|\int \mu(\xi - \eta)\hat{f}(\xi)\,\hat{\bar{g}}(\xi)d\xi\|_{\infty,\eta}\|g\|_{p'} \le \|\psi\|_1 {}_p\|T\|,$$

by part d) above.

f) We will give the proof only for the case $n = 2, k = 1$, and even restrict attention to $\xi_1 = (0,1)$. Higher dimensional cases are very similar, except they involve a lot more notation, while more general cases in $\mathbb{R}^2$ can be obtained by rotations and translations of this one.

In our special case, $\mathbb{R}^{1_0} = \{(-,\lambda)\}$, and we must show that for almost all λ, $\mu(\xi_1, \lambda)$ is a multiplier of $L^p(\mathbb{R}^1)$. Since this λ business could cause real friction with previous notation, we'll prove that $\mu(\xi_1, \xi_2)$ is a multiplier of $L^p(\mathbb{R}^1)$ for almost all ξ_2.

The first pain is that μ is only measurable, and the restriction of μ to one-dimensional hyperplanes is not very well defined. We will use another standard trick in multiplier theory. Assume first that μ is continuous. Let f_i, g_i, $i = 1, 2$ be chosen with $\|f_i\|_p = \|g_i\|_{p'} = 1$; let $f(x_1, x_2) = f_1(x_1)f(x_2)$, and similarly for g. Then:

$$\int \left[\int \mu(\xi_1, \xi_2)\hat{f}_1(\xi_1)\hat{g}(\xi_1)d\xi_1 \right] \hat{f}_2(\xi_2)\hat{g}(\xi_2)d\xi_2 = \int \mu(\xi)\hat{f}(\xi)\hat{g}(\xi)d\xi$$

$$= \int \widehat{Tf}\hat{g} = \int Tfg \le \|Tf\|_p\|g\|_{p'} \le {}_p\|T\|.$$

Define

$$\sigma(\xi_2) = \int \mu(\xi_1, \xi_2)\hat{f}_1(\xi_1)\hat{g}_1(\xi_1)d\xi_1.$$

Rewriting the above computation gives us:

$$\int \sigma(\xi_2)\hat{f}(\xi_2)\hat{g}_2(\xi_2)d\xi_2 \le {}_p\|T\|.$$

Therefore, σ is a multiplier of $L^p(\mathbb{R}^1)$. But Lemma 1.16 told us that $\|\sigma(\xi_2)\|_\infty \le {}_p\|T\|$. Since μ is continuous, σ is continuous, and $\sigma(\xi_2) \le {}_p\|T\|$, for all ξ_2. Rewriting,

$$\left| \int \mu(\xi_1, \xi_2)\hat{f}_1(\xi_1)\hat{g}(\xi_1)d\xi_1 \right| \le {}_p\|T\|$$

$$\left| \int \mu_0(\xi_1)\hat{f}_1(\xi_1)\hat{g}(\xi_1)d\xi_1 \right| \le {}_p\|T\|$$

$$\left| \int T_0 f_1 g_1 \right| \le {}_p\|T\|$$

$$_p\|T_0\| \le {}_p\|T\|.$$

We assumed μ was continuous; we'll use a standard trick to replace continuity assumptions by almost everywhere conditions. Let S denote the unit square, and let $\psi_\delta(\xi) = \delta^{-2}\chi_S(\delta^{-1}\xi)$. Then $\psi_\delta \in L^1$ and $\mu \in L^\infty$, so that $\psi_\delta \star \mu$ is continuous, whence, for every ξ_2,

$$\int \psi_\delta \star \mu(\xi_1, \xi_2)\hat{f}_1(\xi_1)\hat{g}_1(\xi_1)d\xi_1 \le {}_p\|T\psi_\delta\|$$

$$\le {}_p\|T\|\|\psi_\delta\|_1 = {}_p\|T\|.$$

Now, for almost every ξ_2, it happens that almost every ξ_1 is a Lebesgue point of μ_0. These ξ_2 are the ones referred to in the theorem; fix one. The definition of Lebesgue point ξ_1 is that

$$\lim_{\delta \to 0} \psi_\delta \star \mu(\xi_1, \xi_2) = \lim_{\delta \to 0} \delta^{-2} \int_{-\delta}^{\delta} \int_{-\delta}^{\delta} \mu(\eta_1 - \xi_1, \eta_2 - \xi_2)d\eta_1 d\eta_2$$

19

$$= \mu(\xi_1, \xi_2).$$

For this choice of ξ_2, $\psi_\delta \star \mu(\xi_1, \xi_2)$ converges almost everywhere in ξ_1 to $\mu(\xi_1, \xi_2) = \mu_0(\xi_1)$. Of course $(\psi_\delta \star \mu)\hat{f}_1 \hat{g}_1$ also converges; to get the integrals to converge, I need dominated convergence. But

$$|(\psi_\delta \star \mu)\hat{f}_1 \hat{g}_1| \leq \|\psi_\delta \star \mu\|_\infty |\hat{f}_1||\hat{g}_1|$$

$$\leq \|\psi_\delta\|_1 \|\mu\|_\infty |\hat{f}_1 \hat{g}_1|.$$

Therefore,

$$\int \mu_0(\xi_1) \hat{f}_1(\xi_1) \hat{g}(\xi_1) d\xi_1 = \lim_{\delta \to 0} \int \psi_\delta \star \mu \hat{f}_1 \hat{g}_1 \leq {}_p\|T\|.$$

and ${}_p\|T_0\| \leq {}_p\|T\|$ for almost all ξ_2.

1.24 REMARKS: These results are important because they let us deform multipliers into things we can handle. For example, if ψ is in $\mathcal{S}$, and μ is a multiplier, then $\psi\mu$ and $\psi \star \mu$ are multipliers. We already used this in the proof of the above theorem: we changed an arbitrary multiplier into a continuous function, with no side-effects. In general, μ is just in L^∞, and K is some distribution, but if we alter by a compactly supported function ψ, then K_ψ, the inverse Fourier transform of $\psi\mu$, is a continuous function. In the following, changing an awful distribution into a nice regular function is going to be our favorite trick.

The real importance is philosophical. The L^p boundedness of (T, K, μ) is influenced by the size of K, which reflects the smoothness of $\check{\mu} = K$. But the computation of μ is a global average of all of μ against some exponentials, so it is difficult to understand how a local property of μ can affect the Fourier transform K. This is the point of having around $\psi\mu$ or $\psi \star \mu$. The first changes μ into a compactly supported function; the second changes μ into a smoother function.

1.25 THEOREM. *Let $\mu \in L^\infty(\mathbb{R}^n)$ be continuous, and let*

$$S_R f(\theta) = \sum \mu\left(\frac{n}{R}\right) \hat{f}(n) e^{2\pi i n \theta}$$

If the operators S_R are uniformly bounded on $L^p(\mathbb{T}^n), R > 0$, then μ is a multiplier of $L^p(\mathbb{R}^n)$ with operator T, and

$$_p\|T\| \le \sup_R {}_p\|S_R\|.$$

PROOF: We first have to be able to get from functions on $L^p(\mathbb{T}^n)$ to functions of $L^p(\mathbb{R}^n)$, and then transfer operators. If f and g are smooth compactly supported functions on $\mathbb{R}^n$, notice that $f_R(x) = f(Rx)$ is supported in the unit cube in $\mathbb{R}^n$ if R is large enough, and therefore it may be viewed as living on the n-torus. For later use, we compute the effect of this on L^p norms:

$$\|f_R\|_p \, \|g_R\|_{p'} = R^{-n} \|f\|_p \, \|g\|_{p'}$$

To go from Fourier series to Fourier integrals, we use the old idea that a Fourier integral is just a Fourier series with a very large period; the key here is that

$$R^{-n} \sum \mu\left(\frac{n}{R}\right) \hat{f}\left(\frac{n}{R}\right) \hat{g}\left(\frac{n}{R}\right)$$

is a Riemann sum approximation to $\int \mu(\xi)\hat{f}(\xi)\hat{g}(\xi)d\xi$ All we have to do is arrange parameters to get this. Here is the arrangement:

$$\int S_R f_R(\theta) g_R(\theta) d\theta \le {}_p\|S_R\| \, \|f_R\|_p \, \|g_R\|_{p'}$$

and therefore

$$R^n \int S_R(f_R) g_R \le {}_p\|S_R\| \, \|f\|_p \, \|g\|_{p'};$$

$$R^n \int S_R(f_R) g_R = R^n \sum \mu\left(\frac{n}{R}\right) \hat{f}_R(n) \hat{g}_R(n);$$

on the other hand we can compute $\hat{f}\left(\frac{n}{R}\right)$ in terms of Fourier integrals as being

$$\hat{f}_R(n) = R^{-n} \hat{f}\left(\frac{n}{R}\right).$$

We get that

$$R^{-n} \sum \mu\left(\frac{n}{R}\right) \hat{f}\left(\frac{n}{R}\right) \hat{g}\left(\frac{n}{R}\right) \le {}_p\|S_R\| \, \|f\|_p \, \|g\|_{p'}$$

Given that the left-hand side converges,

$$\int \mu(\xi)\hat{f}(\xi)\hat{g}(\xi)d\xi \le {}_p\|S_R\| \, \|f\|_p \, \|g\|_{p'},$$

and

$$\int Tfg \le \sup_p \|S_R\| \, \|f\|_p \, \|g\|_{p'}.$$

This implies that T is bounded on L^p with norm no greater than $_p\|S_R\|$.

1.26 THEOREM. *Let μ be continuous and let T be a multiplier of L^p. Then*

$$Sf(\theta) = \sum \mu(n)\hat{f}(n)e^{2\pi in\theta}$$

is a bounded operator on $L^p(\mathbf{T}^n)$.

1.27 LEMMA. *Let $\omega_\delta(y) = e^{-\pi\delta|y|^2}$. If f is continuous and periodic, then*

$$\lim_{\epsilon \to 0} \epsilon^{\frac{n}{2}} \int f(x)e^{-\epsilon\pi|x|^2}dx = \int f(x)dx$$

1.28 LEMMA. *Let μ, T, S be as in Theorem 1.26. Let P and Q be trigonometric polynomials. If $\alpha + \beta = 1$, then*

$$\lim_{\epsilon \to 0} \epsilon^{\frac{n}{2}} \int T(P\omega_{\epsilon\alpha})\overline{[Q(x)\omega_{\epsilon\beta}]}(x)dx = \int (SP)(x)\bar{Q}(x)dx.$$

PROOF OF LEMMA 1.27: Since the result is linear in f, we can assume that f is a trigonometric polynomial, and we may then assume that $f(x) = e^{2\pi imx}$. The case $m = 0$ is trivial, so we may assume that $\int f(x)dx = 0$. But then

$$\int e^{2\pi imx}e^{-\epsilon\pi|x|^2}dx = \epsilon^{-\frac{n}{2}}e^{-\pi\frac{|m|^2}{\epsilon}},$$

hence

$$\lim_{\epsilon \to 0} \epsilon^{\frac{n}{2}} \int f(x)e^{-\pi\epsilon|x|^2}dx = 0.$$

PROOF OF LEMMA 1.28: Since the result is linear, we can take

$$P(x) = e^{2\pi imx}; \quad q(x) = e^{2\pi ikx}.$$

Then:

$$\epsilon^{\frac{n}{2}} \int T(P\omega_{\epsilon\alpha})\overline{[Q(x)\omega_{\epsilon\beta}]}(x)dx$$

$$= \epsilon^{\frac{n}{2}} \int \mu(\dot{\xi})(P\omega_{\epsilon\alpha})(\xi)\overline{[Q(x)\omega_{\epsilon\beta}]}(\xi)d\xi$$

$$= \epsilon^{\frac{n}{2}} \int \mu(\xi)e^{-\pi\frac{|\xi-m|^2}{\epsilon\alpha}}(\epsilon\alpha)^{-\frac{n}{2}}e^{-\pi\frac{|\xi-k|^2}{\epsilon\beta}}(\epsilon\beta)^{-\frac{n}{2}}d\xi$$

If $m \neq k$, this limit may be computed as zero, and we conclude that

$$\int S(P)\bar{Q} = \int \hat{\mu}(m)e^{2\pi i(\eta-k)\theta}d\theta = 0.$$

If $m = k$, the limit is

$$\lim_{\epsilon \to 0}(\epsilon\alpha\beta)^{-\frac{n}{2}} \int \mu(\xi)e^{-\pi|\xi-m|^2(\frac{1}{\alpha}+\frac{1}{\beta})\frac{1}{\epsilon}}d\xi.$$

22

Since we rigged things so that $\alpha + \beta = 1$, $\frac{1}{\alpha} + \frac{1}{\beta} = \frac{1}{\alpha\beta}$, so that the limit is $\mu(m)$. But

$$\int S(P)\bar{Q} = \int \mu(m)e^0 d\theta = \mu(m).$$

PROOF OF THEOREM 1.26: Trig polynomials are dense in $L^p(\mathbf{T}^n)$, so that we may use them to compute operator norms. Then

$$\int S(P)\bar{Q} = \lim_{\epsilon \to 0} \epsilon^{\frac{n}{2}} \int T(P\omega_{\epsilon\alpha})\bar{Q}\omega_{\epsilon\beta}$$

$$\leq \limsup {}_p\|T\|\epsilon^{\frac{n}{2}}\|P\omega_{\epsilon\alpha}\|_p \, \|Q\omega_{\epsilon\beta}\|_{p'}$$

$$= \limsup {}_p\|T\| \left(\epsilon^{\frac{n}{2}} \int |P(x)|^p e^{-\pi p\alpha|x|^2} \right)^{\frac{1}{p}} \left(\epsilon^{\frac{n}{2}} \int |Q|^{p'} e^{-\pi p'\beta|x|^2} \right)^{\frac{1}{p'}}.$$

Now we get to choose $\alpha = \frac{1}{p}$, $\beta = \frac{1}{p'}$; then Lemma 1.27 tells us that

$$\int S(P)\bar{Q} \leq {}_p\|T\| \left(\int |P|^p \right)^{\frac{1}{p}} \left(\int |Q|^{p'} \right)^{\frac{1}{p'}}.$$

1.29 THEOREM. Let μ be continuous, and let S_R, T be as above. Then $S_R(f)$ converges in $L^p(\mathbf{T}^n)$ for all f in $L^p(\mathbf{T}^n)$ if and only if T is bounded on $L^p(\mathbb{R}^n)$.

PROOF: The uniform boundedness theorem was exactly designed to tell us that convergence of a process on a big set implies uniformity of boundedness of operators. In this case, assume that $S_R(f)$ always converges. Then the operators $f \to S_R(f)$ are uniformly bounded on L^p. By the previous lemmata, T is bounded on $L^p(\mathbb{R}^n)$.

Conversely, assume that T is bounded on $L^p(\mathbb{R}^n)$. Then T_ϵ, defined as in Theorem 1.23c, is also bounded with norm ${}_p\|T\|$; hence the operators T_ϵ are uniformly bounded on $L^p(\mathbf{T}^n)$ by Theorem 1.26. The convergence of $S_R(f)$ in $L^p(\mathbf{T}^n)$ will now be made to follow from convergence on a dense set. We let P be a trig polynomial;

$$\lim_{R \to \infty} S_R P(\theta) = \lim_{R \to \infty} \sum \mu(\frac{n}{R})a_n e^{2\pi i n\theta}$$

$$= \mu(0) \sum a_n e^{2\pi i n\theta)} = \mu(0)P(\theta).$$

1.30 REMARK: This completes the analysis of convergence results; we are supposed to be thinking of the case that $\mu(\xi) = 1$ when $|\xi| < 1$ and $\mu(\xi) = 0$ when $|\xi| \geq 1$. Then S_R represents the Rth partial sum of the Fourier series of f. The L^p convergence of Fourier series is equivalent to the L^p boundedness of a Fourier multiplier. We shall study this boundedness in the next few chapters.

SECTION 1.4   REMARKS FOR CHAPTER 1

1.1): Traditionally Fourier multiplier operators are defined in a different manner. One first analyzes linear operators T bounded on L^p which commute with translations: $\tau(y)T = T\tau(y)$, and one proves that such T are given by convolution with a distribuition. One then proves that T is also bounded on $L^{p'}$, hence on L^2, and is given by a multiplication on the Fourier transform side. An advantage of this approach is that one does not need to assume at the beginning that $\mu \in L^\infty$, as we did. This approach is exploited systematically in Stein and Weiss, [50], Chapter I Section 3.

This raises the question of why operators which commute with translation should be given by a convolution. The reason is that translation itself is an operator which commutes with translation, since $\mathbb{R}^n$ is an abelian group. If we define the finite measure δ_y by $\delta_y(f) = f(y)$, then

$$(\delta_y \star f)(x) = \delta_y(\tau_x \tilde{f})(y) = \tilde{f}(y - x) = f(x - y).$$

So translation is actually given by convolution with a finite measure. I can get more such operators by taking sums and limits of sums of translations. A limit of a sum of translations might happen to converge to an integral; for example $\int K(y)\delta_y dy$, a distribution which acts on functions as $\int K(x - y)f(y)dy = (K \star f)(x)$. The moral is that convolution is nothing more than a weighted average of translations. This is an important intuitive idea that recurs throught this book; a prime example is averaging against the characteristic function of an interval. For example, in proving the fundamental theorem of calculus, one has to compute the limit of

$$\frac{1}{h} \int_x^{x+h} f(y)dy.$$

If I denotes the interval $[0, h]$, then a change of variables shows that this is the same as $\frac{1}{h} \int \chi_I(x - y)f(y)dy$. So the fundamental theorem of calculus really analyzes a limit of convolution operators.

The problem with this philosophy is there's no reason to expect the Fourier transform to enter the picture at all. The simplest place to think about these problems is on $L^2(\mathbf{T}^n)$. The translations are a commuting family of bounded self-adjoint operators on L^2, so the spectral theorem says that they can be simultaneously diagonalized. Since

$$\tau_\psi\left(e^{2\pi in\theta}\right) = e^{2\pi in(\theta - \psi)}$$
$$= e^{-2\pi in\psi}e^{2\pi in\theta},$$

it follows that $\mathbb{C}e^{2\pi in\theta}$ are simultaneous eigenspaces of all the translations. Since the exponentials are complete in L^2, this is the decomposition of L^2 into eigenspaces.

To develop this line of reasoning further, the spectral theorem recovers a function f in L^2 from its projections onto eigenspaces. The projection onto a one dimensional eigenspace generated by a unit vector Φ is given in general by $\langle f, \Phi \rangle \Phi$, or, in this case, by

$$\left[\int f(\theta)\overline{\left(e^{2\pi in\theta}\right)}d\theta\right]e^{2\pi in\theta} = \hat{f}(n)e^{2\pi in\theta}.$$

24

This means that the sum of the projections converges to f in L^2; that is,

$$\sum \hat{f}(n)e^{2\pi in\theta} \to f$$

in L^2. The spectral theorem "explains" the convergence of the Fourier series to the original function.

Still in this abstract context, take an operator T which is bounded on L^2 and commutes with translations. Then T preserves the eigenspaces of the translation operators, so T is also diagonalized by the exponentials. A diagonal operator is determined by its eigenvalues; these are given by the formula

$$T\left(e^{2\pi in\theta}\right) = \mu(n)e^{2\pi in\theta}.$$

The action of T on a general function f can be given in terms of the diagonalization as

$$T(f)(\theta) = \sum \mu(n)e^{2\pi in\theta}$$

and T is now revealed as a Fourier multiplier operator. We'd expect the L^2 norm of T to be the largest eigenvalue of T, that is, the largest of the numbers $|\mu(n)|$; in short: $L_2\|T\| = \|\mu\|_\infty$.

1.12): For a proof of the three lines theorem, see Stein and Weiss, [50], Chapter V Lemma 1.4

1.13): This result is due to M. Riesz; the proof here to Thorin and Tamarkin and Zygmund, [57]. The result is usually called the Riesz-Thorin interpolation theorem. It can be extended to maps from L^{p_i} to L^{q_i}, but with the restriction that $p_i \leq q_i$.

1.19): The weak L^p spaces are special cases of Lorentz spaces; instead of requiring uniform bounds on the growth of the distribution functions, one imposes integrability conditions. A great deal of the theory developed for L^p and weak L^p spaces can be done in the context of Lorentz spaces. See Stein and Weiss, [50] Chapter V, or Hunt [29] for a careful discussion.

1.21): This result is due to Marcinkiewicz, [37]. Like the Riesz-Thorin theorem, it can be extended to maps from L^{p_i} to L^{q_i}. Unlike the Riesz-Thorin theorem, the restriction that $p_i \leq q_i$ is unnecessary.

1.23): This result is due to K. de Leeuw [19]; the proof here is due to Jodeit [30]. The whole result is a lot clearer on $\mathbf{T}^n$; thus,

$$(T_0 f)(\theta_1) = \sum_{n_1} \mu(n_1, n_2)e^{2\pi in\theta_1}\hat{f}(\theta_1) =$$

$$\sum_{n_1}\sum_{n_2} \mu(n_1, n_2)e^{2\pi in\theta_1}\hat{f}(n_1)\left[e^{2\pi in_2\theta_2}\right](n_2)$$

$$= T\left[f(\theta_1)e^{2\pi in_2\theta_2}\right].$$

Since the torus has measure 1, it is easy to relate L^p norms of functions of one variable to those of functions of two variables.

1.25), 1.26): These results appear first in the work of Stein and Weiss; see [50] Chapter VII. Note that the results as stated do not apply to the multipliers that arise from summability of Fourier series; in that case the multiplier μ is the characteristic function of a set. A more careful version shows that they apply to multipliers for which every lattice point in $\mathbb{Z}^n$ is a Lebesgue point.

This whole section brings up a question: why bother looking at the convergence of S_R in L^p? After all, what I really want to know about is the convergence of $S_R f(\theta)$ to $f(\theta)$ for each θ. What does L^p have to do with reality?

First, the example in Section 0 of a continuous function whose Fourier series fails to converge at a point showed that pointwise convergence is too much to hope for. Georg Cantor developed his theory of countable and uncountable sets to construct continuous functions whose Fourier series diverge on more and more complicated sets. The basic modern result in the theory is this: given a set of measure zero, there is a function in L^1 whose Fourier series diverges on that set.

There are two paths one may take. The first path is to analyze the minimal regularity required of a function so that its Fourier series converges at a point. A great deal of important work has been done on this problem; the answers are complex, as are answers to the analagous question: what conditions on a sequence of real numbers guarentees that the associated series converges? As students of convergence tests know, there is no fine line between convergent and divergent series.

We choose a different path: since almost everywhere convergence is the best we can do, what size conditions on f guarentee that the Fourier series of f converges to f almost everywhere? This has a suprisingly simple set of answers. The first result is due to Calderon (see Zygmund [57]) and to Stein [46]. If $1 \le p \le 2$, and $S_R f(\theta)$ converges to f for almost every θ, for all f in $L^p(\mathbf{T}^n)$, then the operator $S^* f(\theta) = \sup_R |f(\theta)|$, is weak (p, p). This is the sort of thing you would hope functional analysis would do for you. You know almost everywhere convergence holds for some dense set of functions; to extend to all functions should require some uniform estimates of something. Since the convergence we want is pointwise, the right uniformity is not uniform boundedness of operators on L^p, but boundedness in L^p of some pointwise-uniform operator, S^*.

But this has several consequences. If convergence occurs for all functions in L^{p_0} for some $p_0 < 2$, then pointwise convergence happens for all p with $p_0 \le p \le \infty$, since all these L^p are contained in L^{p_0}. By the Marcinkiewicz interpolation theorem, S^* is strongly bounded in L^p for $p_0 < p < 2$. But $|S_R f(\theta)| \le S^* f(\theta)$, so that the operators S_R are also uniformly bounded on this range of p. This type of analysis was an essential component of C. Fefferman's proof of divergence of multiple Fourier series.

The moral is this: L^p results are not only helpful in getting pointwise results, they are essential. For a survey of other areas besides Fourier series in which similar results are true, see the article of Gilbert [24].

CHAPTER 2 THE HILBERT TRANSFORM

This chapter is the longest and most important in the book. We discuss only one topic, the one-dimensional theory of summability of Fourier series. What we will be doing is developing the basic models of how (T, K, μ) interact. This means we need to study the relations between a function and its Fourier transform.

SECTION 2.1 INTRODUCTION

We begin by analyzing a class of Fourier multipliers, the Bochner-Riesz means. Their virtue is that they provide a precisely controlled but varying amount of smoothness. They form a model for studying how smoothness affects the Fourier transform.

2.1 DEFINITION. If $\lambda > -1$, define

$$\mu_\lambda(\xi) = \left(1 - |\xi|^2\right)^\lambda_+ .$$

That is, if $|\xi| < 1$, then $\mu_\lambda(\xi) = \left(1 - |\xi|^2\right)^\lambda$; if $|\xi| \geq 1$, then $\mu_\lambda(\xi) = 0$. We also define $K_\lambda = \check{\mu}_\lambda$ and $T_\lambda f = K_\lambda \star f$.

The triple $(K_\lambda, T_\lambda, \mu_\lambda)$ are called Bochner Riesz Means of order λ.

2.2 REMARKS: If $0 \leq \lambda \leq 1$ then μ_λ is Holder continuous of order λ, that is,

$$|\mu_\lambda(r) - \mu_\lambda(s)| \leq C|r - s|^\lambda,$$

and Holder continuity of order λ_0 fails for any $\lambda_0 < \lambda$. If $\lambda \geq 1$, then μ_λ has $[\lambda]$ derivatives, and these are Holder continuous of order $\lambda - [\lambda]$.

The point here is that the multipliers μ_λ have a varying but controlled smoothness. Moreover, if $\lambda = 0$, then μ_λ is the characteristic function of the unit ball in $\mathbb{R}^n$, and therefore the work in Chapter I tells us that the L^p boundedness of μ_0 is equivalent to the L^p convergence of Fourier series.

2.3 PROPOSITION. The convolution kernel $K_\lambda(x)$ is given by

$$c_\lambda \frac{J_{\frac{1}{2}+\lambda}(2\pi|x|)}{|x|^{\frac{1}{2}+\lambda}}$$

$$|K_\lambda(x)| \leq C|x|^{-(1+\lambda)}.$$

2.4 REMARK: The symbol J_ν denotes the Bessel function of order ν, that is, it denotes the solution of the differential equation

$$z^2 D^2 J_\nu + z D J_\nu + (z^2 - \nu^2) J_\nu = 0$$

where we are interested in the solution which is analytic at $z = 0$.

With this identification of the Bessel function, the proposition is easy to prove; we simply compute that $z^\nu \int_{-1}^{+1} e^{izt}(1 - t^2)^{\nu - \frac{1}{2}} dt$ satisfies the differential equation and is analytic. We can get an estimate on the size of $|K_\lambda|$ by changing contours in the integral and doing an asymptotic expansion. In fact, we will do exactly that, in detail, in Chapter 6. We do not perform the computation now because the purpose of this part of the book is to see how the smoothness of μ_λ affects the size of K_λ. The Proposition in question tells us exactly: K_λ's rate of decrease at infinity is directly proportional to the smoothness λ of the multiplier μ_λ. The problem with these Bessel function formulae is that they are global: the explicit formula for K_λ is obtained by averaging all of μ_λ. But μ_λ is a function which is smooth except for $|x| = 1$; since we are trying to understand the effect of the singularity at $|x| = 1$, it is conceptually wrong to average μ_λ over all of $[-1, +1]$. We really would rather localize near the singularity of μ_λ at $|x| = 1$.

2.5 An Intuitive Computation of Fourier Transforms

We want to rethink the entire question of computing Fourier transforms; the question we want to examine is how the smoothness of a function affects the rate of decease of its Fourier transform. What we intend to do is understand some simple examples first, and then expand that to a general understanding.

We have two basic models of the smoothness-decrease problem. The first is a computation done in the analysis of the Hausdorff-Young theorem. We computed there that $\widehat{f\left(\frac{x}{\delta}\right)}(\xi) = \delta \hat{f}(\delta\xi)$. Why is this a result about smoothness? Because as δ becomes smaller, the support of $f\left(\frac{x}{\delta}\right)$ becomes smaller, whilst the size of f is the same. Shrinking δ forces a large change in the size of f over a short range: the derivative is increased. The effect on the Fourier tansform is to multiply by a factor of δ(small), and to dilate $\hat{f}$ by a factor δ. If we think of $\hat{f}$ as being compactly supported, $\hat{f}(\delta\xi)$ has its support increased as δ gets small. This means that the Fourier transform does not decrease as rapidly as when δ was large.

All this is qualitative; the quantitative analysis is contained in the inequalities

$$|x^2 \check{\psi}(x)| \le \|x^2 \check{\psi}\|_\infty \le C\|D_\xi^2 \psi\|_1.$$

There is a technical problem in applying this here; for $0 \le \lambda < 1$, the function μ_λ has no derivatives at all. We are forced to fall back on our inner resources. We will develop a standard model of how to treat singularities such as this. We will sneak up on the singularity of μ_λ; we write μ_λ as a sum of standard functions each of whose singularity we can control. Let $\mu_\lambda = \sum \mu_\lambda \phi_j$, where each ϕ_j is smooth and supported away from $|\xi| = 1$. This is rigged up so that $\mu_\lambda \phi_j$ is smooth, and we can use standard smoothness estimates to find the size of its Fourier transform. Of course we have to add up all the individual Fourier transforms, and as $j \to \infty$ we expect that $\mu_\lambda \phi_j$ is becoming less and less smooth, so that the Fourier transforms are getting larger and larger: potentially, we have a series that isn't going to sum. To get out of this, we need to control the variation of $\mu_\lambda \phi_j$ over its support. If the support is in $[a_j, a_{j+1}]$, we need to see that the size of μ_λ does not change significantly over the length of the interval. Thus, we need constants c_1, c_2 for which

$$c_1 \le \frac{\mu_\lambda(a_{j+1})}{\mu_\lambda(a_j)} \le c_2 \ .$$

28

For a_j close to the singularity at 1, this suggests

$$c_1 \leq \left| \frac{(1 - a_{j+1})}{(1 - a_j)} \right| \leq c_2$$

and $a_j = 1 - 2^{-j}$ will do nicely.

We'll be sloppy and pretend that the singularity is at 0 instead of 1, so we make the ϕ_j all dilates of one function(exploiting the intuition of how dilations increase the size of the derivatives). Let $\phi_j(\xi) = \phi(2^j \xi)$, where we pretend the support of ϕ_j is contained in $I_j \cup -I_j$; here $I_j = \left(1 - 2^{-j}, 1 - 2^{-(j+1)}\right)$. Then $\mu_\lambda \phi_j$ has size approximately $2^{-\lambda j}$, so that approximately,

$$\widehat{\mu_\lambda} = \sum_{j=0}^{\infty} 2^{-\lambda j} \phi_j \check{\mu}_\lambda(x) = \sum_{j=0}^{\infty} 2^{-\lambda j} \check{\phi}_j(x) = \sum_{j=0}^{\infty} 2^{-\lambda j} 2^{-j} \check{\phi}(2^{-j} x).$$

Approximately, then,

$$|K_\lambda(x)| \leq \sum_{j=0}^{\infty} 2^{-(1+\lambda j)} |\check{\phi}(2^{-j} x)|.$$

If we choose ϕ to be very smooth, we can easily control its size; for $\phi \in \mathcal{S}, |\check{\phi}(y)| \leq C$ if $|y| \leq 1$, and $|\check{\phi}(y)| \leq C|y|^{-2}$ if $|y| \geq 1$. Using this accurate estimate,

$$|K_\lambda(x)| \leq C_1 \sum_{|2^{-j}x| \leq 1} 2^{-(1+\lambda)j} + C_2 \sum_{|2^{-j}x| \geq 1} 2^{-(1+\lambda)j} |2^{-j}x|^{-2}$$

$$\leq C_1 \sum_{[\log |x|]-1}^{\infty} 2^{-(1+\lambda)j} + C_2 \sum_{0}^{[\log |x|]+1} 2^{-(1-\lambda)j} |x|^{-2}$$

$$\leq C_1 |x|^{-(1+\lambda)} + C_2 |x|^{-(\lambda+1)} |x|^{-2}.$$

This proves the estimate if $0 \leq \lambda < 1$; if $\lambda \geq 1$, we need to use estimates on $\check{\phi}$ that exploit faster decrease at infinity.

2.6 A Precise Computation of Fourier Transforms

The first time we were sloppy in the computation of Fourier transforms in 2.5 was in the estimate $\mu_\lambda \phi_j = 2^{-\lambda j}$ on the set I_j. The second time we were sloppy was in assuming that the ϕ_j were all dilates of one ϕ; we need to replace dilation intuition by more accurate estimates like

$$|x^2 \check{\psi}(x)| \leq \|x^2 \check{\psi}\|_\infty \leq C \|D_\xi^2 \psi\|_1.$$

We begin with a precise construction of the ϕ_j. Let $\psi(x) = \exp(-x^{-2})$; ψ is C^∞ and supported in $[0, \infty)$. Then the function $\psi(x)\psi(1 - x)$ is smooth and supported in

29

$[0, 1]$. An appropriate dilate and translate of this function, call it ϕ, is supported in $\left[-\frac{5}{4}, -\frac{1}{4}\right]$. Let $\phi_j(\xi) = \phi\left(2^j(|\xi| - 1)\right)$. Notice that if I_j is defined to be the interval $\left[1 - 52^{-(j+2)}, 1 - 2^{(j+2)}\right]$, then the support of ϕ_j is contained in $J_j = I_j \cup -I_j$. Now we set

$$\chi_j(\xi) = \frac{\phi_j(\xi)}{\sum_{k=0}^{\infty} \phi_k(\xi)}$$

if $\xi \in J_j$, while $\chi_j(\xi) = 0$ outside of J_j. Notice that $\left|D_\xi^\alpha \chi_j(\xi)\right| \leq C_\alpha 2^{\alpha j}$ if $\xi \in J_j$.

One of the obligations of precision is that I have to check a lot of petty details; for example that $0 < \sum \phi_k < \infty$ for $\xi \in J_j$. Of these two, the first inequality is trivial; $\phi_k \geq 0$; $\sum \phi_k \geq \phi_j$. But on the interior of $J_j, \phi_j > 0$, while the endpoints of J_j are in the interior of J_{j-1} or J_{j+1}. The sum is positive. Moreover, the only intersections of J_j are with J_{j-1} or J_{j+1}, so that for a fixed ξ, the sum has at most three non-zero terms, making it very finite.

Finally, we will let $\psi_j = \mu_\lambda \phi_j$. By construction, $\psi_j \in \mathcal{S}$; support $\psi_j \subset J_j$, so that we get $\sum \psi_j = \mu_\lambda$. Now to estimate $\check{\psi}_j$, we use the fact that on J_j, μ_λ is smooth, so that

$$|D_\xi^\alpha \mu_\lambda(\xi)| \, \chi_{J_j}(\xi) \leq C_\alpha 2^{-(\lambda-\alpha)j}.$$

Then

$$|\check{\psi}_j(x)| \leq \|\check{\psi}_j\|_\infty \leq \|\phi_j\|_1 = \int_{J_j} |\mu_\lambda| \chi_{J_j} \leq C 2^{\lambda j} |J_j|$$

$$= C 2^{-(1+\lambda)j} |x^2 \check{\psi}_j(x)| \leq \|x^2 \check{\psi}_j\|_\infty \leq \|D_\xi^2 \phi_j\|_1$$

$$\leq C\|D_\xi^2 \mu_\lambda \chi_{J_j}\|_1 + C\|D_\xi \mu_\lambda D_\xi \chi_{J_j}\|_1 + \|\mu_\lambda D_\xi^2 \chi_{J_j}\|_1$$

$$\leq C|J_j| \left(2^{-(\lambda-2)j} + 2^{-(\lambda-1)j} 2^j + 2^{-\lambda j} 2^{2j}\right).$$

In all, $|\check{\phi}_j(x)| \leq 2^{-\lambda j} |2^{-j} x|^{-2}$.

Ok, now the proof is exactly the same as in 2.5. Once again, if $\lambda \geq 1$, the estimates we want just don't happen: $\sum_{j=0} 2^{(1-\lambda)j} |x|^{-2}$ contributes terms of order $|x|^{-2}$, and this simply isn't good enough if $1 - \lambda \leq 0$. To counter this, we can estimate $|x|^3 |\check{\psi}_j|$ and $D_\xi^3(\mu_\lambda \chi_{J_j})$.

2.7 THE FOURIER TRANSFORM AND THE STANDARD MODEL

The word "analysis" may be broken up into its Greek components: "$\alpha\nu\alpha$" and "$\lambda\upsilon\sigma\iota\sigma$", words meaning "up" and "break". In Western intellectual history, analysis is the process of breaking up a process into smaller parts, and attempting to understand each of the parts. There are two hopes here; first, that the individual parts will be amenable to study, though the whole may not be. Secondly, one hopes that the whole can be somehow recaptured from a study of the parts. This latter hope is often a problem; notice that rather than recover complete information about $\check{\mu}_\lambda$, we have only gotten information about the size of $\check{\mu}_\lambda$ near infinity.

What we want to do in this section is make the whole analysis process more intuitively appealing. We saw that isolating the singularity at $|\xi| = 1$ had a lot of advantages; we made μ_λ behave as though it were constant over small intervals. As a consequence, to control the Fourier transform of $\mu_\lambda = \sum \mu_\lambda \chi_{I_j} = \sum 2^{-\lambda j} \chi_{I_j}$ we have only to control $|\check{\chi}_{I_j}|$. Now we are only interested in the size of $|\check{\chi}_{I_j}|$; this means that we can translate I_j to be centered at the origin, and only change the Fourier transform by a phase factor of absolute value 1. Then all the χ_{I_j} are truly dilates of the characteristic function of the interval $I = [-1,1]$. But dilation has a known effect on the Fourier transform; $|\check{\chi}_{I_j}(x)| \leq 2^{-j}|\check{\chi}_I(2^{-j}x)|$. At first we needed to understand the Fourier transform of a whole collection of complicated functions; now all we need to do is understand the Fourier transform of the characteristic function of a single interval.

Intuitively,

$$\check{\chi}_I(x) = \int_I e^{2\pi i x \xi} d\xi$$

is an average of an oscillating exponential over an interval. If the frequency is large, the oscillation is fast, and the average over the entire interval is near zero. The only frequencies which can contribute to $\check{\chi}_I$ must be adapted to the size of the interval: we must have a frequency equal to the length of the interval. We must have $|2^{-j}x| \leq 1$. For such x,

$$|\check{\mu}_\lambda(x)| = 2^{-(1+\lambda)j} \leq |x|^{-(1+\lambda)}.$$

The bad news is that this computation is wrong; if the frequency of a wave is not an exact multiple of the length of I, the Fourier transform will not average to zero over the interval. The error will be of size of one wavelength of the exponential; that is $|2^{-j}x|^{-1}$. If x is large, this gives an error in $|\check{\mu}_\lambda(x)|$ of $\sum_{|2^{-j}x|>1} 2^{-(1+\lambda)j}|2^{-j}x|^{-1}$, which is of order $|x|^{-1}$ for large x. This estimate is lousy.

We can get better estimates by replacing χ_{I_j} with a smooth function ψ with support in I_j, and then we dilate and translate. The same reasoning applies as before, except that if a frequency does not fit exactly into the interval I_j, there is an error in the computation proportional to one wavelength of the exponential, times the height of the the function ψ at the endpoints of the interval. Since ψ vanishes to infinite order near the ends of the interval, the contribution will be very small. If ϕ were linear, the height would be $0 + (slope)(change\ in\ \xi) = (D_\xi\psi)(|wavelength|)$. The total error contributed would be $|wavelength|^2 = |2^{-j}x|^{-2}$; this contributes to the sum a term of size $|x|^{-(1+\lambda)}$, which is what we wanted. If $\lambda \geq 1$, this all falls apart, and we need to use the fact that ψ vanishes to infinite order, so that the height of ψ near an endpoint is like $|D_\xi^k\psi||wavelength|^k$; and so on. We'll do a precise version of this later; the intuition is clear.

There's a basic point here, which in Quantum Mechanics is called the Uncertainty Principle; we cannot simply ignore frequencies which are fast but not perfectly adapted to an interval. These frequencies contribute errors of size proportional to the frequency. If the support of a function ϕ is confined to a small set, then $\check{\phi}$ is not compactly supported; it is real analytic, so definitely not compactly supported. It is not possible to localize both ϕ and $\check{\phi}$ in a small set. Since control of Fourier multipliers requires simultaneous control of μ_λ and $K_\lambda = \check{\mu}_\lambda$, we can expect to work hard for our results.

2.8 COROLLARY. If $\lambda > 0$, T_λ is bounded on $L^p(\mathbb{R}^1)$ for $1 \leq p \leq \infty$.

PROOF: If $\lambda > 0$, $K_\lambda \in L^1$, and therefore

$$_p\|T_\lambda\| \leq \|K_\lambda\|_1 < \infty.$$

31

2.9 LEMMA. *If $\lambda = 0$, T_λ is not bounded on $L^1(\mathbb{R}^1)$.*

PROOF: If $\lambda = 0$, our estimates on T_λ do not show that K_λ is in L^1. But then again, we have lousy estimates. On the other hand, if $T_0 f \in L^1$ for all $f \in \mathcal{S}$, then $\widehat{T_0 f}(\xi)$ is continuous for all ξ. This means that $\chi_{[-1,1]} \hat{f}(\xi)$ is continuous for all those $\hat{f} \in \mathcal{S}$. This is nonsense.

2.10 REMARKS: a) The proof of Lemma 2.9 was almost trivial; it was a lot easier than the analagous result would be for compact torii, $\mathbf{T}^n$. Yet 2.9 and the result on $\mathbf{T}^n$ are equivalent.

b) Does our standard model help to explain the difference between $\lambda > 0$ and $\lambda = 0$? Intuitively, $\mu_\lambda = \sum 2^{-\lambda j} \phi(2^j \xi)$. Since $\phi \in \mathcal{S}$, all the $\phi(2^j \xi)$ give multipliers on L^p with norm equal to $\|\check{\phi}\|_1$. Then we use the really trivial estimate that $_p\|T_\lambda\| \leq \sum 2^{-\lambda j} \|\check{\phi}\|_1 = C \sum 2^{-\lambda j}$. The difference between $\lambda > 0$ and $\lambda = 0$ is dramatically clear. This fails to explain the fact that T_0 will turn out to be bounded on L^p when $p > 1$; this indicates that for $\lambda = 0$ we will need something besides crude estimates to put μ_λ back together from its pieces. We will be able to do it, though.

2.11 DEFINITION. *The Hilbert transform H is the Fourier multiplier operator given on $L^2(\mathbb{R}^1)$ by*

$$(\widehat{Hf})(\xi) = isign(\xi)\hat{f}(\xi).$$

2.12 LEMMA. *T_0 is bounded on L^p if and only if H is bounded on L^p.*

PROOF:: Both operators are characterized by a sharp jump discontinuity; it is not very difficult to make T_0 from H. Notice that:

$$\chi_{(-\infty,0)}(\xi) = \frac{1 + i(isign\xi)}{2}; \; \chi_{(0,\infty)}(\xi) = \frac{1 - i(isign\xi)}{2};$$

$$\chi_{(-1,1)} = \chi_{(-\infty,1)}(\xi)\chi_{(-1,\infty)}(\xi) = \chi_{(-\infty,0)}(\xi - 1)\chi_{(0,\infty)}(\xi + 1).$$

Now Theorem 1.23 shows that these alterations of the Hilbert transform have no significant effect; $_p\|T_0\| \leq \frac{1}{4}(1 +_p \|H\|)^2$.

Conversely, if T_0 is bounded on L^p, then shifting $\chi_{(-1,1)}$ produces $\chi_{(0,2)}$ with no change in multiplier norm. Dilating gives $\chi_{(0,R)}$ and then taking limits over Schwartz functions gives $\chi_{(0,\infty)}$. A similar reasoning applies on the negative side, and then there is no problem adding these to get $isign\xi$.

2.13 REMARK: Replacing T_0 by H seems like an especially stupid thing to do. After all, the characteristic function of the interval is in L^1, so that the convolution kernal associated with it is in L^1. But the multiplier for H is so horrible that the convolution kernel can't possibly be in L^1(otherwise its Fourier transform would decay at infinity). I just replaced a multiplier that was local, whose convolution kernel was integrable, and gotten a multiplier that is global and a kernel not even locally integrable.

2.14 PROPOSITION. *If $f \in S$, then*

$$Hf(x) = \lim_{\epsilon \to 0} \frac{1}{\pi} \int_{\epsilon \leq |y| \leq \frac{1}{\epsilon}} \frac{1}{y}f(x - y)dy.$$

PROOF: We are going to be slightly sloppy with our distributional computations:

$$\lim_\epsilon \frac{1}{\pi} \int_{\epsilon \leq |y| \leq \frac{1}{\epsilon}} \frac{1}{y}e^{-2\pi iy\xi}dy = \lim \frac{2i}{\pi} \int_\epsilon^{\frac{1}{\epsilon}} \frac{1}{y}\sin 2\pi\xi y \; dy$$

$$= \frac{2i}{\pi}sign\xi \lim_\epsilon \int_\epsilon^{\frac{1}{\epsilon}} \frac{1}{y}\sin y \; dy = \left(\frac{2\pi}{\pi}sign\xi\right)\left(\frac{\pi}{2}\right).$$

No real analyst can resist the temptation to evaluate the integral

$$\lim_\epsilon \int_\epsilon^{\frac{1}{\epsilon}} \frac{1}{y}\sin y \; dy$$

without using contour integration. The tricks below were shown us by K. Merryfield; the integrals have all been carefully rigged to converge absolutely, so the formal manipulations we perform are OK.

$$\int_0^\infty \frac{\sin y}{y} dy = \left[\frac{(1 - \cos y)}{2} \right]_0^\infty + \int_0^\infty \frac{1 - \cos x}{x^2} dx$$

But $x^{-2} = \int_0^\infty s e^{-sx} ds$, so that the previous integral is

$$\int_0^\infty \int_0^\infty (1 - \cos x) s e^{-sx} dx ds = \int_0^\infty \left(\frac{1}{s} - \frac{1}{s+1} \right) s ds$$

$$= \int_0^\infty \frac{1}{s^2 + 1} ds = artctan \, \infty - arctan \, 0 = \frac{\pi}{2}$$

2.15 REMARKS: a) Informally, we will write $Hf = \frac{1}{x} \star f$. Of course this convolution converges only for very special f, because of the singularity at $x = 0$. This is what we expected when we replaced T_0 by H. Did we gain anything? The convolution kernel corresponding to K_0 is $\frac{\sin 2\pi x}{x}$, convergent near zero. But neither K_0 nor $\frac{1}{x}$ are in L^1, so that the L^p bounddness properties of these operators doesn't depend on size, but somehow on cancellation between the positive and negative parts of the kernels (compare Remark 1.17; neither the absolute value nor the positive and negative parts of the convolution kernels yield multipliers bounded on L^p for any p at all. Yet without absolute values, we have an operator bounded on L^2 if nothing else).

The conclusion is that cancellation is essential in understanding either T_0 or H. But the respective convolution kernels are $\frac{\sin 2\pi x}{x}$ and $\frac{1}{x}$. The cancellation properties of the Hilbert transform are a lot easier to control.

b) It is easy to do an intuitive computation of the inverse transform of $(sign \xi)$. Since $sign \, \xi = \frac{\xi}{|\xi|}$ is a function invariant under the dilations $\xi \to \delta \xi$, $\delta > 0$, the convolution kernel K must satisfy $\delta K(\delta x) = K(x)$, so that $K(x) = \frac{\omega(\frac{x}{|x|})}{|x|}$, for some function ω which is invariant under dilations. But $sign \, \xi$ is an odd function, and the Fourier transform preserves that, so ω must be a multiple of the odd function invariant under dilations, $\frac{x}{|x|}$. This means that K is $\frac{c}{x}$.

c) We understand that H is not bounded on L^1, but is bounded on L^2. We have proofs of both of these facts, proofs which use Fourier transform techniques. This is not good enough, because the Fourier transform averages too much, and the unboundedness of H on L^1 is due to a purely local discontinuity of the multiplier at $\xi = 0$. We have no local understanding of what the convolution kernel $\frac{1}{x}$ has to do with L^p boundedness; as a distribution it is just as miserable on all functions in any L^p. The next result, and after that the next section of the book, constructs a better understanding of how the convolution kernel influences the L^p boundedness of the operator.

2.16 LEMMA. Let $I = [0, 1]$, and $f = \chi_I$. Then $Hf \in L^p$ if and only if $1 < p < \infty$.

PROOF: To show that Hf is not in L^1, we have to arrange things to avoid cancellation; in the integral $\int_{-1}^1 \frac{1}{x-y} dy$ we want to look at $x - y > 0$. Then if

34

$y \in (-1, 1)$, $x \in (n, n+1)$, $\frac{1}{x-y} \geq \frac{1}{n}$, and therefore

$$Hf(x) \geq \sum \frac{2}{n} \chi_{(n,n+1)}(x) \|Hf\|_1 = \infty.$$

To prove positive results, we'll distinguish between points close to the singularity and points far away: two intervals for x, $(0,2)$ and $(2,\infty)$. In the latter, we use the estimate $\frac{1}{x-y} \leq \frac{1}{n-1}$ on $(n, n+1)$; then $\int_2^\infty \leq 2^p \sum \frac{1}{n^p} < \infty$ for $p > 1$.

On the interval $(0,2)$, we use cancellation to obtain

$$Hf(x) = \log|x+1| - \log|x-1|.$$

The first term is in $L^\infty(0,2)$; the second term can be analyzed symmetrically on $(0,1)$ and $(1,2)$; we will creep up to the singularity to measure its effect. On $(0,1)$, $\int_0^1 |Hf|^p = \sum \int_{1-2^{-k}}^{1-2^{-(k+1)}} |Hf| \leq \sum k^p 2^{-k} < \infty$.

2.17 REMARK: The purpose of computation is insight; what did we learn? First, we got the not-so-suprising result that far from the support of f, Hf behaves like $\frac{1}{x}$. This partly controls the L^p behaviour of Hf. After all, at infinity, $\frac{1}{x}$ is in L^p if and only if $p > 1$. But there is another region, x near the support of f, in which the size of $\frac{1}{x}$ becomes irrelevant, and cancellation becomes important. This region needs the fact that $\int_{\epsilon \leq |y| \leq \frac{1}{\epsilon}} \frac{1}{x} = 0$.

In this section we will prove the $L^2(\mathbb{R}^1)$ bundedness of H without using the Fourier transform. We especially want to see how the cancellation of the kernel $\frac{1}{x}$ is used to control H.

We will fall back on the standard model; we should creep up on the singularity of $isign(\xi)$ at $\xi = 0$ by writing

$$isign(\xi) = \sum 2^{-0j} \phi(2^j \xi).$$

We want an interpretation of this on the convolution kernel side;

$$K(x) = \sum 2^{-j} \check{\phi}(2^{-j}\xi).$$

Intuitively, $\check{\phi}$ has most of its support in $[1, 2]$; this is wrong, but it leads us to suspect that the right way to sneak up on the singularity of the convolution kernel is to write

$$\frac{1}{x} = \sum_{-\infty}^{\infty} K_j(x);$$

here

$$K_j(x) = \frac{1}{x} \chi_{(2^j, 2^{j+1})}(|x|).$$

As expected, $\sum {}_2\|K_j\| = \infty$, so that the L^2 boundedness of H depends on how the K_j interact with each other.

The key intuition that allows us to reconstruct H from the K_j comes from Fourier transform analysis: on the Fourier transform side, the K_j act like multiplication by ϕ_j. The ϕ_j are very like the characteristic function of the interval $\left[2^{-(j+1)}, 2^{-j}\right]$. These are projection operators; moreover they are projections onto orthogonal subspaces of L^2.(I might even think of them as diagonal operators, with 1's along the diagonal). The correct way to measure the size of such an operator is not by adding up the size of all the individual pieces, but by taking the largest of the norms of the pieces, which is finite, in this case.

This discussion is an operator-theoretic version of the Fourier transform proof that H is bounded on L^2. What we have to do is translate this idea about projections into something on the convolution kernel side. Operator theoretically, the condition that we have projections is that the operators T_j corresponding to the kernels K_j satisfy the orthogonality condition $T_k T_j = 0$. Now we have to do two things: i) check that orthogonality is actually true for our operators (they were not defined as projections), and ii) check that orthogonality is enough to give L^2 boundedness.

To check orthogonality, we are not allowed to use the estimate

$${}_2\|T_k T_j\| = \|\hat{K}_k \hat{K}_j\|_\infty,$$

since we are not using the Fourier transform. The only estimate we have relating convolution kernels to operator norms is

$${}_2\|T_k T_j\| = \|K_k \star K_j\|_1,$$

36

so we need to understand the size of $K_j \star K_k$. In the particular case $k = 1, j$ large, we look at Figure 2. The convolution $K_j \star K_k$ should be viewed as an averaging process; in the region $(2^j, 2^{j+1})$, K_j hardly varies at all, while outside this region it is zero. On the other hand, K_1 jumps between positive and negative in an interval of length 2, so we expect it to have zero average against the slowly changing function K_j.

This is our hope; unfortunately the intuition is wrong. There are two problems: K_j is slowly varying, but not constant, so that the positive and negative parts of K_1 see different regions of K_j, and do not average out to zero. The smoothness of K_j controls how large the average really is. Secondly, when $x \in (2^j - 1, 2^j)$, the negative part of K_1 is averaged against 0, while the positive part is averaged against K_j. This gives a quantity of size 2^{-j} on a set of measure 1;

$$\|K_j \star K_1\|_1 > 2^{-j}.$$

The convolution is indeed small, but it is not zero, and our entire philosophical system about projections falls apart.

There is a recovery: a theory of almost orthogonal projections. This theory exploits the fact that $T_j T_k$ is very small; the theory cleverly repeats this estimate to make the interaction between terms very small. The theory uses the spectral radius theorem: the norm of an operator T is given by

$$\|T\| = \lim_{n \to \infty} \|T^n\|^{\frac{1}{n}}.$$

Taking the nth power of T allows me to iterate any estimates I have; on the other hand, I have to keep track of the combinatorics of multiple T_{i_j} averaged against T_{i_k}.

2.18 THEOREM. *Assume that $\{T_k\}$ is a collection of bounded operators on a Hilbert space, and assume that*

$$\|T_k\| \leq C$$

$$\|T_i^* T_j\| \leq \phi^2(i - j)$$

$$\|T_i T_j^*\| \leq \phi^2(i - j).$$

Then

$$\|\sum_{-N}^{N} T_j\| \leq \sum_{-\infty}^{\infty} \phi(j).$$

PROOF: Let T denote $\sum_{-N}^{N} T_j$; then T^*T is self adjoint, and spectral radius applies to it. What we really want to know is the norm of T; the relation is:

$$\|Tf\|^2 = (Tf, Tf) = (f, T^*Tf) \leq \|f\|^2 \|T^*T\|;$$

therefore,

$$\|T\| \leq \|T^*T\|^{\frac{1}{2}}.$$

So we can be content with estimating the norm of this operator. Doing the combinatorics,

$$(T^*T)^n = \sum T_{i_1}^* T_{i_2} T_{i_3}^* \dots T_{i_{2n}}.$$

We expect we can control pairs of iterates, though maybe not the relation of a complicated sum;

$$\|T_{i_1}^* \dots T_{i_{2n}}\| \le \phi^2(i_1 - i_2) \dots \phi^2(i_{2n-1} - i_{2n})$$

$$\|T_{i_1}^* \dots T_{i_{2n}}\| \le C^2 \|T_{i_2} \dots T_{i_{2n-1}}^*\|$$

$$\le \phi^2(i_2 - i_3) \dots \phi^2(i_{2n-2} - i_{2n-1}).$$

The product and then the root of these estimates yields

$$\|T_{i_1}^* \dots T_{i_{2n}}\| \le C\phi(i_1 - i_2) \dots \phi(i_{2n-1} - i_{2n}).$$

Finally,

$$\|(T^*T)^n\| \le C \sum \phi(i_1 - i_2)\phi(i_2 - i_3) \dots \phi(i_{2n-1} - i_{2n})$$

$$\le \left[\sum_{-\infty}^{\infty} \phi(j)\right] \left[\sum \phi(i_1 - i_2)\phi(i_2 - i_3) \dots \phi(i_{2n-2} - i_{2n-1})\right]$$

$$\le \left[\sum_{-\infty}^{\infty} \phi(j)\right]^{2n-2} \left[\sum \phi(i_1 - i_2)\right]$$

$$\le \left[\sum_{-\infty}^{\infty} \phi(j)\right]^{2n-1} (2N).$$

In all,

$$\|(T^*T)^n\|^{\frac{1}{n}} \le \lim_{n \to \infty} [2CN]^{\frac{1}{n}} \left[\sum \Phi(j)\right]^{2 - \frac{1}{n}},$$

and the result follows.

2.19 THEOREM. Let $I_j = (2^j, 2^{j+1})$, $K_j = \frac{1}{x}\chi_{I_j}(|x|)$, $T_j f(x) = K_j \star f(x)$. Then

$$_2\|T_j\| \le 3$$

$$_2\|T_j^* T_i\| = \ _2\|T_j T_i^*\| \le 2^{-|i-j|}.$$

Therefore, the Hilbert transform is bounded on L^2.

PROOF:

$$_2\|T_j\| \le \|K_j\|_1 = \log 4 \le 3.$$

Moreover, recalling that $T_i^* = J_0 J_1 T_i J_0 J_1$, where the $J's$ are the isometries of conjugation and reflection, we get the estimates :

$$_2\|T_j^* T_i\| = \ _2\|T_j T_i^*\|$$

38

$$= {}_2\|T_j T_i\| \le \|K_i \star K_j\|_1.$$

Moreover, if $i > j$, a litttle variable changing shows that

$$K_i \star K_j(x) = 2^{-j} \left(K_{i-j} \star K_0 \right) \left(2^{-j} x \right),$$

and this dilation leaves the L^1 norm of $K_i \star K_j$ unchanged. So it is enough to prove the estimate

$$\|K_m \star K_0\|_1 \le C2^{-m}$$

when m is large.

Our intuitive idea of the convolution is that it represents an average of one function against the other; the computation gets sticky as the different supports of the funtions overlap. We begin by sorting out this overlap: let

$$S_m = \{y \mid 1 < |y| < 2, \ 2^m < |x - y| < 2^{m+1}\}.$$

Then S_m is the set of y for which the integrand in

$$K_m \star K_0(x) = \int K_m(x - y) \, K_0(y) dy$$

is non-zero. Our intuition tells us the integral is small because K_0 has average zero and K_m is slowly varying. We can exploit both facts when we notice that

$$\int_{S_m} \frac{1}{x} \frac{1}{y} dy = 0,$$

and therefore that

$$K_m \star K_0(x) = \int K_m(x - y) K_0(y) dy$$

$$= \int_{S_m} \frac{1}{x - y} \frac{1}{y} dy = \int_{S_m} \left[\frac{1}{x - y} - \frac{1}{x} \right] \frac{1}{y} dy.$$

The difference of the two fractions ought to be small, by the slowly varying trick: we could use Taylor's theorem to estimate it. On the other hand, we can actually do the subtraction, so why not:

$$\left[\frac{1}{x - y} - \frac{1}{x} \right] \frac{1}{y} = \frac{1}{x^2} \left[1 - \frac{y}{x} \right]^{-1}.$$

But if m is large,

$$\left| \frac{y}{x} \right| \le \frac{2}{2^m} < \frac{1}{2},$$

and therefore

$$|K_m \star K_0(x)| \le 2 \frac{1}{x^2} \int_{S_m} dy$$

$$\le \frac{4}{x^2},$$

since $|S_m| \leq 2$. Finally, we have to compute the L^1 norm of $K_m \star K_0(x)$. The support is

$$\{x| \text{ there exists } y \text{ with } 1 < |y| < 2, \ 2^m < |x - y| < 2^{m+1}\}$$

$$\subset \{x| \ 2^m - 2 < |x| < 2^m + 2\},$$

whence

$$\|K_m \star K_0\|_1 < 4 \int_S \frac{1}{x^2} dx \leq C 2^{-m}.$$

We now apply Theorem 2.18, using $\phi(j) = 2^{-|j|}$. We see that

$$S_N f(x) = \sum_{-N}^{N} T_j f(x)$$

are uniformly bounded on L^2. Thus,

$$\lim_{N \to \infty} S_N f(x) = \lim_{\epsilon \to \infty} \int_{\epsilon < |y| < \frac{1}{\epsilon}} \frac{1}{x - y} f(y) dy$$

converges in L^2. On the Schwartz functions, the convergence is to Hf, and therefore

$$_2\|H\| \leq \sup \ _2\|\sum_{-N}^{N} T_j\| < \infty.$$

2.20 REMARK: It is important to see why this worked. In earlier intuitions, we'd gotten simple estimates like

$$Hf \sim \frac{1}{x};$$

here, for some reason, we got

$$Hf \sim \frac{1}{x^2}.$$

Why the improvement? It came from two sources: we computed $K_m \star K_0$, which is like taking $H(K_0)$, since m is so big. Then we used the smoothness of K_m and the fact that $\int K_0 = 0$ to do our estimates. An explicit version of this computation would run like this. Let f be supported on an interval I with center y_0; assume $\int f = 0$. Let $\tilde{I}$ be the interval concentric with I but three times as long. If x is not in $\tilde{I}$, and $y \in I$, then:

$$Hf(x) = \int \frac{1}{x - y} f(y) dy = \int \left[\frac{1}{x - y} - \frac{1}{x - y_0}\right] f(y) dy$$

$$= \int_I \frac{y - y_0}{(x - y)(x - y_0)} f(y) dy.$$

But

$$\frac{1}{x - y} \leq \frac{C}{x}; \quad \frac{|y - y_0|}{|x - y_0|} \leq \frac{C|I|}{|x|},$$

so that

$$|Hf(x)| \leq \frac{C|I|}{|x|} \int_I |f|.$$

2.21 An Attack on the L^p Boundedness of the Hilbert Transform

We know that H is not bounded on L^1 and is bounded on L^2. Moreover our L^1 counterexample gives $Hf \sim \frac{1}{x}$, which sort of suggests that H is weak (1,1). To prove L^p boundedness, then, it would be enough to interpolate between weak L^1 and L^2.

To carry out this program, I need to know how L^1 differs from L^2. The proof of the Marcinkiewicz interpolation suggests a splitting of $f \in L^1$. Let

$$G = \{x| \, |f(x)| \le \frac{\lambda}{2}\}; \quad B = \{x| \, |f(x)| > \frac{\lambda}{2}\}; \quad g = \chi_G, \, etc.$$

Then g is really good:

$$\lambda^2 |\{|Hg| > \lambda\}| \le \int_{\{|Hg|>\lambda\}} |Hg|^2$$

$$\le \|Hg\|_2^2 \le \|g\|_2^2 = \int_G |f|^2 < \lambda \int_G |f|$$

$$\le \lambda \|f\|_1;$$

this gives the weak L^1 behaviour of H on the good part of f. To control H on the bad part, we bring together two of our observations: if f is constant on an interval, we can control Hf; if f has average zero on an interval, we can control Hf. Of course f is not supported on an interval, but we can force it to be. Notice that if f is in S, say, then B is an open set, hence it can be written as a disjoint union of intervals I. Now let f_I denote the average of f over I:

$$f_I = \frac{1}{|I|} \int_I f.$$

Then

$$f\chi_I = [f_I + (f - f_I)] \chi_I = f_I \chi_I + b_I;$$

now $f_I \chi_I$ is constant on each I, and b_I has average zero over I. All in all, then,

$$b = f\chi_B = \sum b_I + \sum f_I \chi_I.$$

We have the decomposition, we can control the Hilbert transform on the pieces; is there anything to stop us from finishing the proof? Yes:

1) There is control of Hb_I off of $\tilde{I}$; there is no control on $\tilde{I}$.
2) The Minkowski inequality does not hold; we can't simply write

$$|\{\sum |f_I| > \lambda\}| \le C \sum |\{|f_I|\} > \lambda\}|;$$

the constant C will depend on how many terms there are in the sum.

The first difficulty means we have to give up all hope of controlling Hb on $B^* = \cup \tilde{I}$. This is not such a big problem:

$$|\{x \in B^* | |Hb(x)| > \lambda\}|$$

41

$$\leq |B^{\star}| \leq \sum |\tilde{I}| = 3 \sum |I|$$

$$= 3|B| = 3|\{|f| > \frac{\lambda}{2}\}|$$

$$\leq \frac{6}{\lambda} \int_{\{|f| > \frac{\lambda}{2}\}} |f|$$

$$\leq 3\frac{\lambda}{\lambda} |\{|f| > \frac{\lambda}{2}\}|$$

$$\leq \frac{6}{\lambda} \|f\|_1.$$

This gives the weak L^1 estimate of b on $B^{\star}$. The second difficulty is harder to handle; we will be forced to do our estimates on some space where the triangle inequality does hold; otherwise we have no hope of reconstructing Hf from its pieces. Now, b is the L^1 part of f, and we expect that Hb will act like $\frac{1}{x^2}$, L^1 ought to be the right space for controlling Hb:

$$\lambda |\{x \in B^{\star} \mid |H(\sum b_I)| < \lambda\}| \leq \int_{(B^{\star})^c} |H(\sum b_I)|$$

$$\leq \sum \int_{(B^{\star})^c} |H(b_I)|.$$

We can't hope to control the action of H on the intersections of the $\tilde{I}$, and we can't control $H(b_I)$ particularly well except on $(\tilde{I})^c$. This means that we shall have to increase our integrals; $(B^{\star})^c \subset (\tilde{I})^c$, and:

$$\sum \int_{(B^{\star})^c} |H(b_I)| \leq \sum \int_{(\tilde{I})^c} |H(b_I)|.$$

Now we analyze each term individually. If y_I is the center of I, we have

$$(Hb_I)(x) = \int_I \left[\frac{1}{x-y} - \frac{1}{x-y_I} \right] b_I(y) dy$$

$$\int_{(\tilde{I})^c} |H(b_I)(x)dx| \leq \int_{(\tilde{I})^c} \int_I \left| \frac{1}{x-y} - \frac{1}{x-y_I} \right| |b_I(y)| dy dx$$

$$= \int_I |b_I(y)| \int_{(\tilde{I})^c} \left| \frac{1}{x-y} - \frac{1}{x-y_I} \right| dx dy.$$

But the inner integral is something that can be computed explicitly; it is bounded by

$$4 \log \left| 1 - \frac{|y-y_I|}{|I|} \right|.$$

Since

$$y \in I, \ x \in (\tilde{I})^c,$$

42

$$\frac{|y - y_I|}{|I|} \le \frac{1}{2},$$

and we have a bound

$$\int_{(\tilde{I})^c} |H(b_I)(x)dx| \le 8 \int_I |b_I(y)|.$$

Then

$$\sum \int_{(B^*)^c} |H(b_I)| \le 8 \sum \int_I |b_I(y)| dy$$

$$\le 8 \sum \int_I (|f(y)| + |f_I(y)|) \, dy = 8 \sum \int_I |f(y)| dy + 8 \sum |I||f_I|$$

$$= 16 \sum \int_I |f(y)| dy = 16 \int_B |f(y)| dy \le 16\|f\|_1.$$

That handles *that* term. Now the only term remaining is the constant term, $f_I \chi_I$. Surely handling the constant term can be left to the mentally deficient. As before, we need a space where the triangle inequality holds; but as before our intuition tells us that $H(f)(x) \sim \frac{1}{x}$, so that L^1 is not the right space to work on. L^2 is satisfactory; thus,

$$\left|\left\{\left|H\left(\sum f_I \chi_I\right)\right| > \lambda\right\}\right|$$

$$\le frac1\lambda^2 \int \left|H\left(\sum f_I \chi_I\right)\right|^2$$

$$\le C\frac{1}{\lambda^2} \int \left|\sum f_I \chi_I\right|^2$$

$$\le C\frac{1}{\lambda^2} \sum \int_I |f_I \chi_I|^2;$$

since the I's are disjoint. To finish the proof, we need to get rid of one power of λ and one power of $|f_I|$: we need the estimate

$$\frac{1}{|I|} \int_I |f| \le C\lambda$$

i.e.

$$|f|_I \le C\lambda$$

this would give us the final estimate

$$C\frac{1}{\lambda} \sum \int_I |f_I| = C\frac{1}{\lambda} \sum |I||f_I|$$

$$= C\frac{1}{\lambda} \sum \int_I |f| = C\frac{1}{\lambda} \int_B |f| \le C\frac{1}{\lambda}\|f\|_1.$$

43

And of course if the sum has only one term, we really need the estimate

$$\frac{1}{|I|} \int_I |f| \le C\lambda.$$

Alas, the correct estimates all go the other way: since $I \subset B$,

$$|f_I| \ge \frac{\lambda}{2}.$$

What all this means is that the constant terms $f_I \chi_I$ actually contribute the main problem of bounding H. This ought not to have suprised us; it was they who are like $\frac{1}{x}$ at ∞; it is they who contribute the non-L^1 behaviour.

What can we do? The two inequalities

$$|f_I| \ge \frac{\lambda}{2},$$

which we know, and

$$|f|_I \le C\lambda,$$

which we want, suggest that we have to be careful about choosing the intervals I. They must be closely adapted to the function f; intuitively, they must be chosen so that f does not vary too much over I. Formally, what we need are the following four properties:

We must decompose $\mathbb{R}$ into disjoint sets
$\mathbb{R} = G \cup B$;
$B = \cup I$, I disjoint and satisfying $f_I \le C\lambda$
$|B| \le C\frac{1}{\lambda}\|f\|_1$
If $x \in G, |f(x)| \le \lambda$.

If we can do this for any L^1 function, we can use our bag of tricks to prove that H is weakly bounded on L^1; we can then interpolate and use duality to get it bounded of all L^p between 1 and ∞. What we have gained is the complete elimination of H. The question now is a question about L^1 functions, and whether they can be chopped into pieces which do not vary too much over a small interval. The question is, how much "smoothness", on the average, does an L^1 function have?

How can we control the variation of a function over an interval? Getting $f_I > \lambda$ is the easy part; for an arbitrary interval and an arbitrary λ there is no hope that we can also get $f_I < C\lambda$. We begin by picking our intervals so that $f_I < C\lambda$; all intervals for which this happens are thrown into B; then we worry about the other three properties.

Fix a λ and an f, we might as well assume f is non-negative. The first minor detail is whether there are any good intervals. Since $\int f < \infty$, we can always find J, N such that if

$$|J| \geq N, \quad \frac{1}{|J|} \int_J f < \lambda.$$

On the other hand, we can't take all such intevals, since we intend to bound $|B|$ by $\frac{1}{\lambda}\|f\|_1$. We have a simple but important way to get around this: take an interval J with

$$\frac{1}{|J|} \int_J f < \lambda.$$

Now let I denote the right or left half of J, and assume that

$$\lambda < \frac{1}{|I|} \int_I f.$$

Such an interval is a good one to keep for B, because it has the lower bound properties on averages that we need. Of course the hard part is the upper bound property. Here's where the trick comes in:

$$\frac{1}{|I|} \int_I f < \frac{1}{|I|} \int_J f$$

$$= \frac{2}{|J|} \int_J f < 2\lambda.$$

Thus we have not lost the upper bound on the averages. Now we continue this partitioning process. The left and right-hand sides of the intervals always are disjoint, and if we find a particularly nice half interval I satisfying

$$\lambda < \frac{1}{|I|} \int_I f,$$

we keep it for B. This process produces a collection of disjoint intervals which give averages bounded below by λ and above by 2λ. Because the intervals are disjoint, we even get

$$|B| = \sum |I| \leq \sum \frac{1}{\lambda} \int_I f \leq \frac{1}{\lambda}\|f\|_1.$$

Well, what goes wrong this time? First, we haven't even tried to control B^c. Second, what makes me think that if

$$\frac{1}{|J|} \int_J f < \lambda,$$

I can ever find a subinterval I with

$$\lambda \le \frac{1}{|J|} \int_I f?$$

Of course, nothing stops me from subdividing the two halves of J and subdividing them again, and again. Why should the process ever stop? Well, we do have some choices. If the process never stops, we clearly don't want J for B; on the other hand, anything in G has to satisfy if $x \in G$, then $|f(x)| \le \lambda$. All we know about points $x \in G$ is that there is an infinite sequence of intervals $|I_j|$ containing x, with $|I_{j+1}| = \frac{1}{2}|I_j|$, and we never get $\lambda < f_{I_j}$, that is, we always have $f_{I_j} \le \lambda$. So why should we conclude that $f \le \lambda$?

The question is a basic one: can we recover information about f from information about the averages of f? If f is a continuous function, the answer is simple:

$$\lim_{h \to 0} \frac{1}{h} \int_x^{x+h} f(y)dy = f(x),$$

since this is simply a differentiation of the integral. If we let $I_j = [x, x + 2^{-j}]$, we get that

$$\lim_{j \to \infty} f_{I_j} = \lim_{j \to \infty} \frac{1}{I_j} \int_{I_j} f = f(x)$$

and $f_{I_j} \le \lambda$ implies that $f(x) \le \lambda$.

Again, there are technical difficulties using this. First, I_j is not the interval $[x, x + 2^{-j}]$; we have no control over I_j other than $x \in I_j$. So we cannot use a direct differentiation of the integral argument.

Secondly, it is well known that the derivative of the integral depends on the function. Darboux's theorem tells me that the derivative of a function satisfies the intermediate value property. This means that if f skips a value, there is no differentiable F with $F' = f$. We cannot always recover f from its averages.

To get the correct differentiation of integrals, we must ignore what happens on sets of zero measure, and ask only for convergence almost everywhere: does

$$\lim_{\substack{|I| \to 0 \\ x \in I}} f_I = f(x)$$

for almost all x? If f is continuously differentiable, this is an easy ϵ-δ argument. The problem, as usual, is going from the dense set to the whole of L^1. This is a standard problem, and it has a standard solution: take f_n to be a sequence of smooth functions converging in L^1 to f. What I want to do is say that

$$f(x) = \lim_n f_n(x) = \lim_n \lim_I (f_n)_I = \lim_I \lim_n (f_n)_I = \lim_I f_I$$

So what I really want to do is interchange two limits, but I can never do this without some uniform convergence. We had the same problem in summing Fourier series; we had to get a uniform estimate on operator norms. So too here, except that this is a pointwise operator, so we need pointwise uniformity; we need an estimate on

$$\sup_{\substack{I \\ x \in I}} \frac{1}{|I|} \int_I f.$$

46

2.22 DEFINITION. *Let f be locally integrable. The Hardy-Littlewood maximal function of f is denoted by f^* and is defined as*

$$f^*(x) = \sup_{\substack{I \\ x \in I}} \frac{1}{|I|} \int_I |f(y)| dy$$

2.23 LEMMA. *Assume $f \in L^1$ and f is not equal to 0 a.e.. Then f^* is not in L^1.*

PROOF: Choose N so that $\int_N^N \geq \frac{1}{2}\|f\|_1$. Now choose any x with $|x| \geq N$. Let $I = [-N, N]$;

$$f^*(x) \geq \frac{1}{|I|} \int_I |f|$$

$$= \frac{1}{2x} \int_{-|x|}^{|x|} |f| \geq \frac{1}{2x} \int_{-N}^{N} |f|$$

$$\geq \frac{1}{4x}\|f\|_1.$$

Thus for large x,

$$f^*(x) \geq \frac{C}{x}, \quad C = \|f\|_1 \geq 0.$$

2.24 THEOREM. *The map $f \to f^*$ is weak (1, 1).*

PROOF: We can assume f is non-negative. Let

$$E = \{x \mid f^*(x) > \lambda\}.$$

We have an actual philosophy of how to bound the measure of E, to wit: for each $x \in E$, there is an I with $x \in I$ and $f_I > \lambda$. Then for any $x \in I$, $f^* > \lambda$, so that $x \in E$. This means that E is covered by a bunch of intervals. If they were all disjoint, we'd get

$$|E| = |\cup I| = \sum |I|$$

$$\leq \sum \frac{1}{\lambda} \int_I f = \frac{1}{\lambda} \int_E f \leq \frac{1}{\lambda}\|f\|_1.$$

The bad part is that they way we've defined them, there is no way that the intervals I are going to be disjoint.

Our strategy is to discard most of the intervals, but to retain enough that we don't throw away most of the mass of the set. The ability to do this is called a covering lemma, and it has nothing to do with functions in L^1 or anything except the geometry of intervals and how they intersect.

In deciding which intervals to keep and which to throw away, it turns out that we really have little choice. If we keep an interval I_1, we have to throw out all the intervals which intersect it. How can we be sure that the intervals we threw away didn't have most of the measure of E? There is a simple trick: if we have two overlapping intervals, we will always keep the largest, call it I_*. Say we had 20 intervals which used to intersect I_*, and we threw all of them away. Have we thrown away most of the measure of E? No! If $x \in I \cap I_*$, then $I \subset (x - |I|, x + |I|) \subset (x - |I_*|, x + |I_*|)$ since I_* is the largest of the intervals. Since $x \in I_*$, the whole of $(x - |I_*|, x + |I_*|)$ is contained in an interval concentric with I_*, with length 5 times as large. In throwing away all the intervals that intersected I_*, we decreased the collective measure by a factor of at most 5, and this is a fixed factor independent of the number of intervals, the size of I_*, or anything. Our trick is going to work. There are technical details to work on, of course.

We begin by assuming that E has finite measure, and then by fixing, for each $x \in E$, an I with $x \in I$ and $f_I > \lambda$. This fixed collection is denoted by $\mathcal{B}_\infty$. We won't get a largest element in $\mathcal{B}_\infty$, but we can get close. Let $L_1 = \sup\{|I| \mid I \in \mathcal{B}_\infty\}$. If $L_1 = \infty$, choose I with $|I| > |E|$, and we are done. Otherwise, choose an I_1 with

$$|I_1| > \frac{1}{2}L_1; \ \mathcal{B}_2 = \{I \in \mathcal{B}_2 \mid I \cap I_1 = \phi\},$$

$$L_2 = \sup\{|I| \mid I \in \mathcal{B}_2\}$$

etc. Of course if $\sum |I_j| = \infty$, we are done again; otherwise we have to do some work.

We never did try to show that $\cup I \subset \cup I_j$; we'd need that if $x \in E$, $x \in I$ for some I, and we must find a k so that $x \in I_k$. What does this strategy mean in terms of the $\mathcal{B}$'s? We need to find a k such that $I \in \mathcal{B}_k$ but I is not in $\mathcal{B}_{k+1}$. This is not too difficult, because the lengths of intervals in the $\mathcal{B}$'s gets smaller. Assume $I \in \mathcal{B}_j$ for all j; then $\frac{1}{2}|I| < \frac{1}{2}L_j < |I_j|$. On the other hand, $\sum |I_j| < \infty$, so that either $|I| = 0$ or the sum terminates. But the sum can terminate only if $L_j = 0$, and again $|I| = 0$. This shows that there is a k such that $I \in \mathcal{B}_k$ but I is not in $\mathcal{B}_{k+1}$.

This is good for us, because $I \in \mathcal{B}_k$ implies that

$$\frac{1}{2}|I| \le \frac{1}{2}L_k \le |I_k|,$$

and if I is not in $\mathcal{B}_{k+1}$ there is an $x \in I \cap I_k$. As in the intuitive introduction, $I \subset I_k^*$. Thus, for every $I \in \mathcal{B}_1$, there exists a k with $I \subset I_k^*$. Then $E \subset \cup I_k^*$ and

$$|E| \le |\cup I_k^*| \le \sum |I_k^*|$$

$$= 5 \sum |I_j|$$

and $|\{f^* > \lambda\}| \le \frac{5}{\lambda}\|f\|_1$.

2.25 COROLLARY. If $1 < p \le \infty$, then $\|f^*\|_p \le C_p\|f\|_p$.

PROOF: Note that

$$|f_I| \le \frac{1}{|I|} \int_I |f| \le \|f\|_\infty,$$

so that $f^*(x) \le \|f\|_\infty$ and therefore

$$\|f^*\|_\infty \le \|f\|_\infty.$$

We may use the Marcinkiewicz interpolation theorem, 1.21, to get the result. There is a minor point: we only proved the Marcinkiewicz theorem for linear operators, but the proof given carries over to the case of sublinear operators, those satisfying $T(f + g) \le T(f) + T(g)$.

2.26 COROLLARY. *If $f \in L^1$, then for almost every x,*

$$\lim_{\substack{|I| \to 0 \\ x \in I}} \frac{1}{|I|} \int_I f(y) dy = f(x).$$

PROOF: If f is continuous then this is trivial. To go from the dense set to all of L^1, we use the uniformity of convergence. The proof will be a little peculiar since the maximal operator is not in fact bounded on L^1. Let

$$Tf(x) = \left| \limsup_I f_I(x) - \liminf_I f_I(x) \right|.$$

If g is continuous, $Tg = 0$; we take any $f \in L^1$ and any continuous g and note that

$$Tf = T(f - g + g) \leq T(f - g) + Tg = T(f - g) \leq 2(f - g)^*.$$

Then

$$|\{x | Tf(x) > \lambda\}| \leq |\{x | (f-g)^*(x) > \frac{\lambda}{2}\}| \leq \frac{10}{\lambda} \|f - g\|_1.$$

Since g is arbitrary and the left hand side does not contain g,

$$|\{x | \ Tf(x) > \lambda\}| = 0.$$

This means that $Tf = 0$ for almost all x.

2.27 COROLLARY. *If $p < 1$ and I is any interval, then*

$$\|f^* \chi_I\|_p \leq C_p |I|^{\frac{1}{p} - 1} \|f\|_1.$$

PROOF:

$$\int_I |f^*|^p = p \int_0^\infty \lambda^{p-1} |I \cap \{f^* \geq \lambda\}| d\lambda$$

$$\leq p \int_0^\infty \lambda^{p-1} \min\{|I|, \frac{c}{\lambda} \|f\|_1\} d\lambda$$

$$= p \int_0^{c|I|^{-1}\|f\|_1} |I| \lambda^{p-1} d\lambda + cp \int_{c|I|^{-1}\|f\|_1}^\infty \lambda^{p-1} \lambda^{-1} \|f\|_1 d\lambda$$

$$= c|I|^{1-p} \|f\|_1^p + \frac{C_p}{1-p} |I|^{1-p} \|f\|_1^p.$$

2.28 THEOREM. *If $f \in L^1$, $f > 0$, and $\lambda \geq 0$ are fixed, then there exist measurable sets G, B and intervals I such that:*
1) $\mathbb{R} = G \cup B; G \cap B = \phi$
2) $B = \cup I$; *the I have disjoint interiors and satisfy $\lambda < f_I \leq 2\lambda$*
3) $|B| \leq \frac{c}{\lambda} \|f\|_1$

4) *for almost all $x \in G$, $|f(x)| \le \lambda$.*

PROOF: We begin as in the introduction to this section, by dividing $\mathbb{R}$ into intervals J of equal length, satisfing

$$|J| \ge \frac{1}{\lambda}\|f\|_1,$$

so that we automatically have

$$f_J = \frac{1}{|J|}\int_J f \le \frac{1}{|J|}\|f\|_1 \le \lambda.$$

We will keep J if $\lambda < f_J$; otherwise we divide J in half, to get J_0, J_1. If $\lambda < f_{J_i}$, put J_i into the collection we keep. If we decide to keep the interval, notice that

$$f_{J_i} = \frac{2}{|J|}\int_{J_i} f$$

$$\le \frac{2}{|J|}\int_J f \le 2\lambda,$$

so that for intervals we keep, $\lambda < f_{J_i} \le 2\lambda$ automatically. Moreover, the intervals we keep have disjoint interiors, since they are subsets of intervals with disjoint interiors. The intervals we keep have union B, and we let G be the complement. The only part of the theorem which is not immediate from the construction is 4). If $x \in G$, x is in some initial J which was not kept, and was also in a subdivision which was not kept, etc. There is a chain of intervals I_k with $x \in I_k$ and with $|I_{k+1}| = \frac{1}{2}|I_k|$. Since no I_k was ever kept for B, $f_{I_k} \le \lambda$, and now Theorem 2.26 tells us that for almost all $x \in G$, $f(x) \le \lambda$.

2.29 REMARK: We would rather have taken $B = \{x | f^*(x) \ge \lambda\}$, and we could have avoided the selection process above. But how could we construct the intervals I? The condition $f_I \le 2\lambda$ suggests that I is associated with g, but in fact $I \cap G = \phi$. What's going on? Assume that $y \in \tilde{I} \cap G$. Then

$$f_I = \frac{3}{|\tilde{I}|}\int_I f \le 3f_{\tilde{I}} \le 3f^*(y) \le 3\lambda.$$

So in fact this sort of bound has very much to do with being in G. What we need in order to make this precise is to know that if $x \in B$, there is an interval I with $x \in I$, and $G \cap \tilde{I} \ne \phi$. One way to rephrase this is that the distance of x to B^c is comparable to the length of I.

2.30 THEOREM. *Let B be a proper open subset of $\mathbb{R}$. Then B can be written as a union of intervals I with disjoint interiors satisfying*

$$|I| \le d(B^c, I) \le 4|I|.$$

PROOF: The easiest trick is to let

$$\mathcal{B}_k = \{I | \, 2^k < |I| < 2^{k+1} \text{ and } 2^k < d(I, B^c) \le 2^{k+1}\}.$$

50

The intervals in $\mathcal{B}_k$ will cover B and they automatically have the right distance property. Of course they're not disjoint, so we have some work to do. We'll start with a standard collection and take halves, just as before. Let

$$\mathcal{D}_j = \{I|\ I = [k2^{-j}, (k+1)2^{-j}]\ for\ some\ k\}.$$

The intervals in $\mathcal{D}_j$ are called dyadic intervals; they automatically have the disjointness property we need. Moreover, the intervals in $\mathcal{D}_{j+1}$ are halves of the intervals in $\mathcal{D}_j$. The disjointness property can be phrased slightly differently: if two intervals I and J have interiors with non-trivial intersection, then one interval is contained in the other.

Now we set

$$\mathcal{B}_j = \{J \in \mathcal{D}_j \mid there\ exists\ x \in J\ with\ 22^j < d(x, B^c) \le 22^{j+1}\}.$$

We claim that if

$$J \in \mathcal{B}_j,\ |J| \le d(J, B^c) \le 4|J|.$$

This should be trivial, since J has length around 2^j and the distance between J and B^c is comparable to the distance betwen x and B^c. In detail, we know there is an x in J with

$$22^j < d(x, B^c) \le 22^{j+1};$$

therefore

$$2|J| < d(x, B^c) \le 4|J|$$

and then

$$d(J, B^c) = \inf\{|z - y| \mid z \in J,\ y \in B^c\}$$

$$\le \inf\{|x - y| \mid y \in B^c\} = d(x, B^c).$$

Similarly,

$$d(J, B^c) = \inf\{|z - y| \mid z \in J,\ y \in B^c\}$$

$$= \inf\{|z - x + x - y| \mid z \in J,\ y \in B^c\}$$

$$\ge \inf\{|x - y| \mid y \in B^c\} - \sup\{|z - x| \mid z \in J\} = d(x, B^c) - |J| \ge 2|J| - |J|.$$

To finish the proof of the theorem, we need to show that B is a union of disjoint intervals, presumably the I's in the $\mathcal{B}_j$. If $J \in \mathcal{B}_j$, $d(J, B^c) > 0$, since B^c is closed, $J \subset B$. On the other hand, if $x \in B$, $d(x, B^c) > 0$ since B is closed. Thus there is a $J \in \mathcal{D}_j$ with $x \in J$, since $\cup \mathcal{D}_j = \mathbb{R}$. To get disjointness, we will use intervals of maximal size and use the dyadic properties. Things get a bit mystical at this point, because many intervals are contained in each other and we tend to count them many times. What we need to show is that if $x \in B$, then x is contained in one and only one interval of maximal size. Now, if an interval J contains x,

$$|J| \le d(J, B^c) \le d(x, B^c),$$

so that the collection of intervals containing x has a bounded size. Moreover, the dyadic intervals containing x are ordered by inclusion, so their union is a dyadic interval of maximal size containing x. The maximality now implies disjointness.

51

2.31 COROLLARY. *Assume $f \in L^1$, $f \geq 0$ and λ are fixed. Then there exist sets G, B, and intervals I with disjoint interiors, satisfying:*

1) $\mathbb{R} = G \cup B$,
2) $B = \cup I$, $f_I \leq c\lambda$,
3) $|B| \leq \frac{c}{\lambda}$,
4) $f(x) \leq \lambda$ *for almost all $x \in G$.*

PROOF: Let

$$G = \{x \mid f^*(x) \leq \lambda\}; \ B = \{x \mid f^*(x) > \lambda\}.$$

Then part 1) is clearly true, and differentiation of the integral gives 4) automatically. Since the maximal operator is weak $(1, 1)$, 3) is also automatic. To get the intervals, notice that B is an open set, so Theorem 2.30 gives us disjoint intervals I having good distance properties. These properties must be translated into the average-type conditions of 2).

Let I be one of our intervals and let $\tilde{I}$ be the interval concentric with I but having ten times the length. Since $d(I, B^c) \leq 4|I|$, there must be an x with $x \in \tilde{I} \cap B^c$. Of course

$$f_{\tilde{I}} \leq f^*(x) \leq \lambda.$$

But then

$$f_I \leq 10 f_{\tilde{I}} \leq 10\lambda.$$

Now we can prove the boundedness of the Hilbert transform. Rather than do so, we abstract out the essentials and prove a more general theorem.

2.32 THEOREM. *Asume (T, K, μ) is a multiplier of L^2, and that K satisfies*

$$\sup_y \int_{|x| > 2|y|} |K(x - y) - K(y)| dx \leq C.$$

Then T is weak $(1, 1)$, and is therefore bounded on L^p for all p with $1 < p < \infty$.

PROOF: All we really need to do is repeat the computation at the end of the last section. Fix $\lambda > 0$, and apply Theorem 2.28 with $f > 0$ and $\frac{\lambda}{2}$. Let

$$g = f\chi_G, \ b = f\chi_B, \ h_I = (b - b_I)\chi_I.$$

Since $|Tf| \leq |Tg| + |Tb|$, we may perform the weak $(1, 1)$ estimates on each piece separately.

$$|\{|Tg| > \frac{\lambda}{2}\}| \leq \frac{4}{\lambda^2} \|Tg\|_2^2$$

$$\leq \frac{C}{\lambda^2} \|g\|_2^2 = \frac{C}{\lambda^2} \int_G |g|^2$$

$$= \frac{C}{\lambda^2} \int_G |f||f| \leq \frac{C}{\lambda} \int_G |f|$$

$$\leq \frac{C}{\lambda} \|f\|_1.$$

To handle the bad part, we let $b = \sum(b - b_I)\chi_I + b_I\chi_I$. Note that the disjointness of the I implies that this sum converges ae. Then

$$\left|\left\{|T\left(\sum b_I\chi_I\right)| > \frac{\lambda}{4}\right\}\right|$$

$$\leq \frac{16}{\lambda^2}\left\|T\left(\sum b_I\chi_I\right)\right\|_2^2$$

$$\leq \frac{C}{\lambda^2}\left\|\sum b_I\chi_I\right\|_2^2$$

$$= \frac{C}{\lambda^2}\int|\sum b_I\chi_I|^2 = \frac{C}{\lambda^2}\sum\int|b_I\chi_I|^2$$

$$= \frac{C}{\lambda^2}\sum\int_I|b_I|^2.$$

The sum has been pulled out from the square because the intervals in the sum are disjoint. That is, pointwise,

$$|\sum b_I\chi_I|^2 = \sum|b_I\chi_I|^2.$$

Now

$$b_I = (f\chi_B)_I = f_I \leq C\lambda,$$

so we again have

$$\frac{C}{\lambda^2}\sum\int_I|b_I|^2 \leq \frac{C}{\lambda}\sum\int_I|b_I|$$

$$= \frac{C}{\lambda}\sum|b_I||I| = \frac{C}{\lambda}\sum\int_I|b|$$

$$= \frac{C}{\lambda}\int_B|f| \leq \frac{C}{\lambda}\|f\|_1.$$

As we know, the troublesome parts of the equation are the L^1 pieces, the h_I. Let $B^* = \cup\tilde{I}$; then

$$\left|\left\{x \in B^* \mid |T(\sum h_I)| > \frac{\lambda}{4}\right\}\right|$$

$$\leq |B^*| \leq 4|B| \leq \frac{C}{\lambda}\|f\|_1.$$

On the other hand,

$$\left|\left\{x \in (B^*)^c \mid |T(\sum h_I)| > \frac{\lambda}{4}\right\}\right|$$

$$\leq \frac{4}{\lambda}\int_{(B^*)^c}|T(\sum h_I)| \leq \frac{4}{\lambda}\sum\int_{(\tilde{I})^c}|T(h_I)|$$

$$= \frac{4}{\lambda}\sum\int_{(\tilde{I})^c}\left|\int K(x - y)h_I(y)dy\right|dx.$$

Now $h_I = (b - b_I)\chi_I$, so that $\int_I h_I = 0$. If y_I denotes the center of I, then we can insert a term $-\int_I K(x - y_I)h_I dy$ at will; this gives

$$\frac{4}{\lambda}\sum \int_{(\tilde{I})^c} \left| \int_I \left[K(x - y) - K(x - y_I) \right] h_I(y) dy \right| dx$$

$$\leq \frac{4}{\lambda}\sum \int_I \int_{(\tilde{I})^c} |K(x - y) - K(x - y_I)| \, dx |h_I(y)| dy$$

$$= \frac{4}{\lambda}\sum \int_I \int_{(\tilde{I}-y_I)^c} |K(z + y - y_I) - K(z)| \, dz |h_I(y)| dy.$$

To analyze this integral, notice that if we let $I = [-L + y_I, L + y_I]$, and if we assume $y \in I$ and $z \in (\tilde{I} - y_I)^c$, then $|z| > 2L > 2|y - y_I|$, and it follows that for such y,

$$\int_{(\tilde{I}-y_I)^c} |K(z + y - y_I) - K(z)| \, dz |h_I(y)| dy$$

$$\leq \int_{|z|>2|y-y_I|} |K(z + y - y_I) - K(z)| \, dz |h_I(y)| dy \leq C$$

by the hypotheses on K. This means that the final term is controlled by

$$\frac{C}{\lambda}\sum \int_I |h_I|$$

$$\leq \frac{C}{\lambda}\sum \int_I |b| + \frac{C}{\lambda}\sum \int_I |b_I|$$

$$\leq \frac{C}{\lambda}\|f\|_1.$$

This completes the proof, as all four terms have been bounded by $\frac{C}{\lambda}\|f\|_1$.

2.33 REMARK: The hypotheses of the theorem apply to the kernels

$$K_\epsilon(x) = \frac{1}{x}\chi_{\{\epsilon<|x|<\frac{1}{\epsilon}\}}(x),$$

and all the estimates in the hypotheses are uniform independent of ϵ. It follows that the operators $T_\epsilon f = K_\epsilon \star f$ are uniformly bounded on L^p, and therefore that

$$\lim_{\epsilon \to 0} K_\epsilon \star f$$

converges in L^p. On a dense set in L^2, it converges to the Hilbert transform, so that $T_\epsilon f$ converges in L^p to Hf. We cannot conclude anything about pointwise convergence; it is the old problem of interchanging limits, and requires some sort of uniform estimate. We need to understand $\sup_\epsilon |K_\epsilon \star f|$.

SECTION 2.5 THE MAXIMAL HILBERT TRANSFORM

In the last section, we showed that the Hilbert transform was bounded on L^p; unfortunately the argument proceeded through an L^p limiting process. The defect with this is that for an arbitrary function in L^p, we do not really know that the Hilbert transform exists pointwise. The purpose of this section is to get pointwise information by controlling the correct maximal function.

2.34 DEFINITION. *The maximal Hilbert transform of a function f is denoted H^*f and is defined as*

$$H^*f(x) = \sup_{\epsilon > 0} \left| \int_{|y| > \epsilon} f(x - y)\frac{1}{y}dy \right|.$$

2.35 REMARK: I want to look at the maximal Hilbert transform and the Hardy-Littlewood maximal functions in a more general context to identify a unifying theme. For f^*, let $\psi(x) = \frac{1}{2}\chi_I(x)$; here $I = [-1, 1]$. If $\psi_\epsilon(x) = \frac{1}{\epsilon}\psi\left(\frac{x}{\epsilon}\right)$, and $I_\epsilon = [-\epsilon, \epsilon]$, then

$$\psi_\epsilon \star f(x) = \frac{1}{2\epsilon} \int_{x-\epsilon}^{x+\epsilon} f(y)dy = \frac{1}{|I_\epsilon|} \int_{I_\epsilon} f(y)dy$$

$$\leq \sup_{\substack{I \\ x \in I}} \frac{1}{|I|} \int_I f(y)dy \leq f^*(x).$$

A similar formula holds for H^*f; let

$$\psi(x) = \frac{1}{x}\chi_{(1,\infty)}(|x|),$$

and note that

$$\psi_\epsilon \star f(x) = \int_{|y| > \epsilon} f(x - y)\frac{1}{y}dy.$$

Again

$$H^*f = \sup_{\epsilon > 0} |\psi_\epsilon \star f|.$$

2.36 DEFINITION. *Assume $\psi \in L^1$, $\psi \geq 0$, $\int \psi = 1$; let*

$$\psi_\epsilon(x) = \frac{1}{\epsilon}\psi\left(\frac{x}{\epsilon}\right).$$

Then ψ_ϵ is called an approximate identity.

2.37 REMARK: The term "approximate identity" comes from a result we will soon prove: if ψ is sufficiently nice, $\psi_\epsilon \star f$ converges to f as $\epsilon \to 0$. Thus ψ_ϵ acts approximately like an identity for convolution.

What we want to do is measure the rate of convergence of $\psi_\epsilon \star f$ to f. Using the fact that $\int \psi_\epsilon = 1$,

$$|\psi_\epsilon \star f(x) - f(x)| = \left| \int \psi_\epsilon(y)[f(x - y) - f(x)]dy \right|.$$

Now say f has some sort of continuity properties; say

$$|f(x - y) - f(x)| \leq C|y|^{\lambda}.$$

Then the integral above is dominated by

$$C \int |\psi_\epsilon(y)||y|^{\lambda} dy = C\epsilon^{\lambda} \int |\psi(y)||y|^{\lambda} dy = C\epsilon^{\lambda}.$$

Thus $\psi_\epsilon \star f$ converges to f, and the rate of convergence is controlled by the smoothness of f. This argument works, say, if $f \in \mathcal{S}$, and we need the usual uniform estimate to get to L^p.

2.38 PROPOSITION. *Assume ψ is even, decreasing, and in L^1. Then there is a constant C, independent of ψ, such that:*

$$\sup_{\epsilon > 0} |\psi_\epsilon \star f(x)| \leq C\|\psi\|_1 f^*(x).$$

PROOF: We will show that for all ψ satisying the hypotheses,

$$|\psi \star f(0)| \leq C\|\psi\|_1 f^*(0).$$

This allows us to pull some cheap tricks. First, we replace f by $\tau_x f$, and we note that

$$|\psi \star f(x)| = |(\tau_x \psi \star f)(0)|$$
$$= |\psi \star [\tau_x f](0)| \leq C\|\psi\|_1 [\tau_x f]^*(0)$$
$$= C\|\psi\|_1 f^*(x).$$

These equalities follow since convolution commutes with translation, and since the estimate

$$|\psi \star f(0)| \leq C\|\psi\|_1 f^*(0)$$

is valid for $\tau_x f$ as well as for f. For our next trick, we notice that if ψ satisfies the hypotheses, so does ψ_ϵ, and the L^1 norms of these two functions are equal. Therefore,

$$|\psi_\epsilon \star f(x)| \leq C\|\psi\|_1 f^*(x).$$

We now just take the supremum of both sides to get the result of the proposition.
To estimate

$$|\psi \star f(0)| = \left| \int \psi(-x)f(x)dx \right| = \left| \int \psi(x)f(x)dx \right|,$$

we remark that ψ is decreasing and integrable, so that

$$\lim_{R \to \infty} \psi(R) = 0.$$

The local integrabity also implies that

$$\lim_{\epsilon \to 0} \int_0^\epsilon f = 0,$$

56

so we may integrate by parts and ignore boundary terms:

$$\left| \int_0^\infty \psi(x)f(x)dx \right| = \left| \int_0^\infty \int_0^y f(x)dx\, d\psi(y) \right|.$$

Since ψ is deceasing, $d\psi \leq 0$, and the last integral is dominated by

$$-\int_0^\infty \int_0^y |f(x)|dx\, d\psi(y) \leq -\int_0^\infty 2y\frac{1}{2y}\int_y^y |f(x)|dx\, d\psi$$

$$\leq -\int_0^\infty 2yf^*(0)d\phi \left[-2\int_0^\infty y\, d\phi \right]$$

$$= 2f^*(0)\int_0^\infty \phi = f^*(0)\|\psi\|_1.$$

2.39 PROPOSITION.
$$H^*f(x) \leq [Hf]^*(x) + Cf^*(x).$$

PROOF: Although we know that

$$H^*f = \sup_{\epsilon>0} |\psi_\epsilon \star f(x)|,$$

ψ_ϵ is not integrable. The trick here is to write it as an integrable piece plus a singular piece, and then to look at the action of the singular part as a Hilbert transform; the integrable piece contributes $Cf^*(x)$, and the singular piece gives us $[Hf]^*(x)$. Now, our intuition about Hilbert transforms is that the integrable piece comes from smoothness of the kernel, and is of size about $\frac{1}{x^2}$; to get this estimate we let ϕ be a smooth, even decreasing function with $\phi \geq 0$, $\int \phi = 1$, supp $\phi \subset [-1,1]$. Let $\varphi(x) = \psi(x) - \phi \star \frac{1}{x}$. Then:

$$|\varphi(x)| = |\psi(x) - \phi \star \frac{1}{x}| = |\psi(x)\int \phi(y)dy - \phi \star \frac{1}{x}|$$

$$\leq \int_{|y|\leq 1} \left| \frac{1}{x} - \frac{1}{x-y} \right| \phi(y)dy$$

$$\leq \frac{1}{x^2}\int_{|y|\leq 1} \left| \frac{y}{1-\frac{y}{x}} \right| \phi(y)dy.$$

If $|x| \geq 2$, we can use the fact that $|y| \leq 1$ to bound the integrand, and obtain

$$|\varphi(x)| \leq \frac{C}{x^2},$$

as we expected. Then

$$\psi = \phi \star \frac{1}{x} + \varphi$$

57

$$\psi_\epsilon = \left(\phi \star \frac{1}{x} \right)_\epsilon + \varphi_\epsilon$$

$$= \phi_\epsilon \star \frac{1}{x} + \varphi_\epsilon$$

and

$$H^* f(x) \leq \sup_\epsilon \left| \phi_\epsilon \star \frac{1}{x} \star f(x) \right| + |\varphi_\epsilon \star f(x)|$$

$$\leq \sup_\epsilon \left[\phi_\epsilon \star Hf(x) \right] + sup_\epsilon |\varphi_\epsilon \star f(x)|.$$

Since ϕ is dominated by an integrable decreasing function, 2.38 applies. We have seen that φ is dominated by an even integrable decreasing function in the range $|x| \geq 2$; if $|x| \leq 2$, we look more closely. In the range $|x| \leq 1$, $\phi(x) = 0$ and then $\varphi = -\phi \star \frac{1}{x}$. Since ϕ is smooth,

$$-\phi \star \frac{1}{x}$$

is continuous and therefore bounded. In the range $1 \leq |x| \leq 2$, ϕ is smooth and so is $\phi \star \frac{1}{x}$. Thus φ is dominated by an even decreasing integrable function, 2.38 applies, and the second term is also dominated by the Hardy-Littlewood maximal function.

2.40 THEOREM. H^* is weak $(1, 1)$.

PROOF: Note first that 2.39 implies that H^* is bounded on L^p for $1 < p < \infty$, since the Hilbert transform and the Hardy-Littlewood maximal functions are bounded there. We cannot conclude that the operator has any boundedness on L^1, since it is a composition of operators neither of which preserve L^1. What we will do is use the techniques of 2.32, the proof of the boundedness of the Hilbert transform. We can use the same old intervals I as before, since they are independent of ϵ. The technical difficulties arise when we estimate terms of the form

$$\int_{I^c} \int_I |\psi_\epsilon(x - y) - \psi_\epsilon(x)| \, dx dy.$$

We have destroyed the smoothness of $\frac{1}{x}$ by chopping it off with ψ_ϵ, so we cannot use the usual smoothness arguments. On the other hand, if I is not near the singularity, we are safe. This means that we will have to adjust our intervals to ϵ.

As in the proof of 2.32, we only have to work hard to get the estimates on $H^* (\sum h_I)$, because all the other estimates refer the problem back to L^2, and both the Hilbert transform and the maximal function are bounded from L^2 to L^2. For the constant terms, we fix an ϵ and break the class of all I into three groups:

a) We look at all those I for which $|x - y| < \epsilon$ for all $y \in I$. Then

$$\int \psi_\epsilon(x - y) h_I(y) dy = \int_{|x-y|>\epsilon} \frac{1}{x - y} h_I(y) \chi_I(y) dy = 0.$$

This, we can handle.

58

b) Those I for which $|x - y| > \epsilon$ for all $y \in I$. These intervals don't see the lack of smoothness that we introduced, so we can estimate:

$$\int_{|x-y|>\epsilon} K(x - y)h_I(y)\chi_I(y)dy = \int_I [K(x - y) - K(x - y_I)]\, h_I(y)dy$$

and the estimates continue as before.

c) Those I for which there is a $y_0 \in I$ with $|x - y_0| = \epsilon$. Then there is a jump from 0 to $\frac{1}{x-y}$ in the convolution kernel. Fortunately, the relation between $|x - y_0|$ and ϵ allow us to control $x - y$ by ϵ, and then $\int_I \frac{1}{x-y}h_I(y)dy$ can be related to an average of h_I over an interval of radius ϵ. This latter is controlled by the Hardy-Littlewood maximal function.

To actually do this proof, we must first relate the condition $x \in \tilde{I}^c$, $y \in I$ to the condition $|x - y| < \epsilon$. We claim that for such x, y,

$$\frac{\epsilon}{2} < |x - y| < \frac{3\epsilon}{2}$$

$$|x - y| \geq |x - y_0| - |y - y_0| \geq \epsilon - |I|.$$

But

$$\epsilon = |x - y_0| \geq 2|I|; \quad -|I| > \frac{-\epsilon}{2}; \quad |x - y| > \frac{\epsilon}{2}.$$

On the other hand, $y \in I$ implies that

$$|x - y| \leq |x - y_0| + |y - y_0| = \epsilon + |y - y_0| \leq \epsilon + |I| < \frac{3\epsilon}{2}.$$

Then:

$$\left| \int_{|x-y|<\epsilon} \frac{1}{x - y}h_I(y)\chi_I(y)dy \right|$$

$$\leq \int_{|x-y|>\epsilon} \frac{1}{|x - y|}|h_I(y)\chi_I(y)|dy$$

$$\leq \frac{2}{\epsilon} \int_{|x-y|>\epsilon} |h_I(y)\chi_I(y)|dy$$

$$\leq \frac{C}{\epsilon} \int_{|x-y|<\frac{3\epsilon}{2}} |h_I(y)|\chi_I(y)dy.$$

This will be enough to finish the proof: for any $\epsilon > 0$,

$$\left| \int_{|x-y|>\epsilon} \frac{1}{x - y} \left(\sum h_I\chi_I \right) dy \right|$$

$$\leq 0 + \sum \int_I |K(x - y) - K(x - y_I)|\, |h_I(y)|dy$$

$$+ \sum \frac{2}{\epsilon} \int_{|x-y|<\frac{3\epsilon}{2}} |h_I(y)|\chi_I(y)dy.$$

Thus, for $x \in (B^*)^c$,

$$|H^*f(x)| \leq \sum \int_I |K(x - y) - K(x - y_I)|\, |h_I(y)|dy + \left[\sum |h_I(y)|\chi_I \right]^*,$$

and this estimate is enough for us to finish the proof.

2.1): Bochner-Riesz means are descended from Riesz means $(1 - |\xi|)_+^\lambda$. The multiplier properties are equivalent, since they differ only by multiplication with a function in S. Altering the multiplier by $|\xi|^2$ as in Bochner [1] allows us to compute precise formulae for the T_λ, and these played a key role in Bochner's proof of the failure of localization.

2.4): Details on Bessel's operator ∇_ν can be found in Watson [55] 6.1. The formula quoted here is due to Sonine; see also Stein and Weiss [50].

2.18): The first version of this is due to Cotlar, whence the phrase "Cotlar's lemma". A strengthened version was obtained by Stein; the proof here is from Knapp-Stein [33]. A more geometric result is in Fefferman [21]. For the spectral radius theorem, see Rudin [43].

2.23): Again the most intuitively appealing way to see this is to take f a δ function. Stein [46] shows how this sort of intuition can be made precise.

2.26): There is a converse to 2.26, using Stein's result on limits of sequences of operators [46]. We have always remarked that a uniform estimate is helpful in interchanging limits; Stein's results show that the convergence of the limits in 2.26 implies the weak (1, 1) estimate.

2.28): The decomposition of $\mathbb{R}$ into good and bad sets is called the Calderon-Zygmund decomposition after [3]. This result and 2.24 extend to $\mathbb{R}^n$ if we use balls or cubes for the averaging.

2.30): This result is due to Whitney [56] and was established to control differentiability properties of functions near the boundary of a set. Why does it help to do so?

2.32): The proof that the Hilbert transform is bounded on L^p is due to M. Riesz [41]. Riesz used techniques from complex variables; in particular, the Hilbert transform is the operator which takes an harmonic function to its conjugate (so that the pair is analytic). Kolmogoroff [34] proved the weak (1, 1) boundedness using complex variables again, establishing the existence of boundary values for functions analytic in the interior of the unit disc. The proof given here is a variant of that given by Calderon and Zygmund; its special virtue is that it extends to several dimensions. We have followed the presentation in Stein [49]. The condition on the kernel is from Hormander [28]. Alternative smoothness conditions are possible; see Stein [49].

2.40): We have stated this result as though it applied to Hormander-type kernels K, but in fact we needed to use pointwise estimates on K, as in the estimate $|K(x - y)| \leq \epsilon$. For a more general approach, see Stein [49].

CHAPTER 3 GOOD λ AND WEIGHTED NORM INEQUALITIES

In the last chapter, we got global estimates of the size of Hf in terms of the size of f. This is not the last word on the subject; the types of proofs we did suggest that there might be estimates of $Hf(x)$ in terms of $f(x)$ or $f^*(x)$. Instead of L^p estimates, we might get pointwise estimates.

We will in fact do something close, but weaker. If I is an interval, can we expect $\int_I |Hf|^p \leq C \int_I |f|^p$? If we imagine I to be a very small interval, this would be close to pointwise information, but it would still let us use function-theory and L^p - space ideas. Our model question for understanding this is the following: if $\omega > 0$, is it true that

$$\int |Hf|^p \omega \leq C \int |f|^p \omega$$

for all f? This is called a "weighted norm inequality"; we may think of ω as carrying most of its mass on an interval, with very small size outside. The technical tool we shall use to analyze this problem is called a good-λ inequality. It compares the size of Hf to the size of f^*, and gives a precise estimate of the intuition that f^* controls the size of Hf.

SECTION 3.1 GOOD λ INEQUALITIES

Our intuition is that when Hf is large, f^* must be large as well. This is expressed precisely by the following theorem, saying that measure of the set where Hf is large and f^* is small can be made small.

3.1 THEOREM. *There is a $C > 0$ such that, for all $\gamma > \frac{1}{2}C$ and for all λ,*

$$|\{x \mid H^*f(x) > 2\lambda, f^* < \gamma\lambda\}| \leq C\gamma \left|\{x \mid |H^*f(x)| > \lambda\}\right|.$$

PROOF: We begin by localizing the problem to intervals; we apply the Whitney decomposition theorem 2.30 to the set $B = \{H^*f > \lambda\}$. We get disjoint intervals I with $B = \cup I$, $|I| \leq d(B^c, I) \leq 4|I|$. Since the intervals are disjoint, we can get by with proving

$$|\{x \in I \mid H^*f(x) > 2\lambda, f^* > \gamma\lambda\}| \leq C\gamma|I|.$$

What is our model for the control of H^*f by f? In controlling terms of type c) in the proof of Theorem 2.40, we had x close to the interval J and h_j supported in J. This gave us control of H^*f by f^*. Here, f is not supported in I, and there will be additional terms for those x not near I. The Whitney decomposition controls this for us: if x is not near I, x is near B^c, and on B^c, $H^*f \leq \lambda$.

This intuition suggests a decomposition of f into a part near I and a part concentrated on $\tilde{I}^c$. We choose two reference points which control the size of f: $z_0 \in I$ with $f^*(z_0) \leq \gamma\lambda$; this will control the local part of H^*f as in 2.40c. We will also choose an $x_0 \in B^c$ and get for free that $H^*f(x_0) \leq \lambda$. Since $d(B^c, I) \leq 4|I|$, we can select x_0 so that $d(x_0, I) \leq 5|I|$, so this will determine what "close to I" and "far away from I" mean. There are technical problems, though, because $H^*f(x)$ is not directly comparable to $H^*f(x_0)$; the latter requires that we measure with intervals centered at x_0, and I is not so centered. We construct an interval J larger that I, but at least centered at x_0. To begin.

Choose $z_0 \in I$ with $f^*(z_0) < \gamma\lambda$; choose $x_0 \in B^c, d(x_0, I) \leq 5|I|$. Choose an interval J centered at x_0 with 20 times the length of I. Clearly $I \subset J$, for if

$$x \in I, \ |x - x_0| \leq d(x_0, I) + |I| \leq 6|I|.$$

Let $g = f\chi_J, \ g = f\chi_{J^c}$. We shall show that

$$\left| \{x \in I \mid H^*g(x) > \frac{\lambda}{2} \} \right| \leq \frac{C}{\lambda}|I|f^*(x_0) \leq C\gamma|I|;$$

$$H^*b(x) \leq H^*f(x_0) + Cf^*(x_0) \leq \lambda + C\gamma\lambda.$$

Given these results,

$$|\{x \in I \mid H^*g(x) > 2\lambda, f^* < \gamma\lambda\}|$$

$$\leq |\{x \in I \mid H^*g(x) > 2\lambda - H^*b, f^* < \gamma\lambda\}|$$

$$\leq |\{x \in I \mid H^*g(x) > 2\lambda - C\gamma\lambda, f^* > \gamma\lambda\}|$$

$$\leq \left| \{x \in I \mid H^*g(x) > \frac{\lambda}{2} \} \right| \leq C\gamma|I|,$$

at least if $\gamma < \frac{1}{2}C$.

The estimate of H^*g on I is not difficult, even though this is supposed to be the worst part of H^*, since we already know H^* is weak $(1, 1)$:

$$\left| \{x \in I \mid H^*g(x) > \frac{\lambda}{2} \} \right| \leq \frac{C}{\lambda}\|g\|_1 = \frac{C}{\lambda} \int_J |f|$$

$$= \frac{C}{\lambda}|J||J|^{-1} \int_J |f| = 20\frac{C}{\lambda}|I|\frac{1}{|J|} \int_J |f|.$$

But $z_0 \in I \subset J$ so that

$$|f|_J \leq f^*(z_0) < \gamma\lambda;$$

the whole term is bounded by $20C\gamma|I|$.

To control $H^*b(x)$ by $H^*f(x_0)$, we try to write

$$\int_{|x-y|>\epsilon} \frac{b(y)}{x - y} dy$$

in terms of

$$\int_{|x_0-y|>\epsilon} \frac{b(y)}{x_0 - y} dy.$$

This introduces two types of errors: those that come from $\frac{1}{x-y} - \frac{1}{x_0-y}$ and those from $\{|x_0 - y| > \epsilon\}\Delta\{|x - y| > \epsilon\}$. To control these, we set

$$S_1 = \{y \mid |x_0 - y| > \epsilon, |x - y| < \epsilon\}$$

62

$$S_2 = \{y \mid |x_0 - y| < \epsilon, |x - y| < \epsilon\}$$

$$E_j = \int_{S_j} \frac{|b(y)|}{|x - y|} dy.$$

Then:

$$\left| \int_{|x-y|>\epsilon} \frac{b(y)}{x - y} dy \right| \le$$

$$\left| \int_{|x_0-y|>\epsilon} \frac{b(y)}{x - y} dy \right| + E_1 + E_2$$

$$\le \left| \int_{|x_0-y|>\epsilon} \frac{b(y)}{x_0 - y} dy \right| + \int_{|x_0-y|>\epsilon} \left| \frac{1}{x_0 - y} - \frac{1}{x - y} \right| dy + E_1 + E_2.$$

The first term, as arranged, is bounded by $H^* b(x_0)$; unfortunately the type of control we have at x_0 is over $H^* f(x_0)$, not $H^* b(x_0)$. This means more work, and shows how our choice of J comes in. If we take $y \in J^c$, then $|y - x_0| > 10|I|$. Thus, if $y \in supp\ b = J^c$,

$$\{y \mid |x_0 - y| > \epsilon, y \in J^c\} = \{y \mid |x_0 - y| > \max\{\epsilon, 10|I|\}\}$$

and

$$\left| \int_{|x_0-y|>\epsilon} \frac{b(y)}{x_0 - y} dy \right| = \left| \int_{\{y \mid |x_0-y|>\max\{\epsilon, 10|I|\}\}} \frac{f(y)}{x_0 - y} \chi_{J^c} dy \right|$$

$$\le \left| \sup_\delta \int_{|x_0-y|>\delta} \frac{f(y)}{x_0 - y} dy \right| = H^* f(x_0).$$

The term controlled by smoothness of the convolution kernel is something we've handled before; it is dominated by

$$\int_{|x_0-y|>\epsilon} |x_0 - y|^{-1} |x - y|^{-1} |x_0 - x| |f(y)| \chi_{J^c} dy.$$

We need to control this term by $f^*(z_0)$, so we need to relate x_0 and z_0, and replace $|x_0 - y|$ with this relation. The fact that no such relation exists merely means that we construct it ourselves. Let

$$F_k = \{y \mid 2^{k+1} \epsilon > |x_0 - y| > 2^k \epsilon\}.$$

We shall show that if $x \in I, y \in J^c, y \in F_k$, then:

$$k \ge \log \left(\frac{10|I|}{\epsilon} \right) = C_\epsilon |x_0 - y|^{-1} > (2^k \epsilon)^{-1}$$

$$|x - y|^{-1} < (2^{k-1} \epsilon)^{-1}$$

$$|z_0 - x_0| < 2^{k+1} \epsilon$$

63

$$|x - x_0| < 6|I|.$$

If we can prove these, then the term controlled by smoothness is dominated by

$$\sum_{k=C_\epsilon}^{\infty} \frac{1}{2^k \epsilon} \int_{F_k} |f(y)| dy \, \frac{|I|}{2^{k-1}\epsilon}$$

$$\sum_{k=C_\epsilon}^{\infty} \frac{10|I|}{\epsilon} 2^{-(k-1)} \frac{1}{2^k \epsilon} \int_{|y-z_0| \leq \frac{1}{2^{k+2}\epsilon}} |f(y)| dy$$

$$\leq \frac{40|I|}{\epsilon} f^*(z_0) 2^{-C_\epsilon} \leq 400 f^*(z_0).$$

The estimates now take nothing but patience; we illustrate a few. First, $y \in F_k$ implies that $2^{k+1}\epsilon > |x_0 - y|$, so that $y \in F_k \cap J^c$ implies $k \geq C_\epsilon$. Since $y \in F_k$,

$$|x_0 - y|^{-1} \leq (2^k \epsilon)^{-1}$$

automatically; to estimate $|x - y|$,

$$|x - y| \geq |x_0 - y| - |x - x_0| > 2^k \epsilon - 6|I|.$$

But $2^{k+1}\epsilon > |x_0 - y| > 10|I|$ so $-6|I| > \frac{\epsilon}{2} 2^k$ and $|x - y| \geq 2^{k-1}\epsilon$. Finally,

$$|z_0 - x_0| \leq d(x_0, I) + |I| \leq 6|I| < 10|I| < 2^{k+1}\epsilon,$$

and

$$|x_0 - x| \leq d(x_0 I) + |I| \leq 6|I|.$$

To finish the proof, we have to control the error terms E_i by $f^*(z_0)$; we need to replace $|x_0 - y|$ information by $|z_0 - y|$ information. For the E_1 term,

$$S_1 = \{y \mid |x - y| < \epsilon, |x_0 - y| > \epsilon\}.$$

Since $z_0 \in I$ and $x \in I$, y should be close to x and the distance from x to z can be at most $|I|$. But this has to be much smaller than ϵ, because $x \in I$ and $y \in J^c$. Thus:

$$\epsilon > |x - y| > 10|I|,$$

$$|z_0 - y| < |z_0 - x| + |x - y| < |I| + \epsilon < 2\epsilon,$$

and

$$S_1 \subset \{y \mid |y - z_0| < 2\epsilon\}.$$

Moreover,

$$|x - y| \geq |x_0 - y| - |x_0 - x| > \epsilon - 6|I|$$

$$\geq \epsilon - \frac{6}{10}\epsilon|x - y|^{-1} > \frac{5}{2\epsilon}$$

and

$$E_1 \leq \frac{5}{2\epsilon} \int_{|y-z_0|<2\epsilon} |f(y)| dy \leq 5 f^*(z_0).$$

In the term from $S_2 = \{y \mid |x_0 - y| < \epsilon, |x - y| > \epsilon\}$, we note that y has to be close to z_0 because x_0 is; again we have to compare $|I|$ to ϵ:

$$\epsilon > |x_0 - y| > |x - y| - |x - x_0| \geq 20|I| - 6|I| = 14|I|;$$

$$|z_0 - y| \leq |z_0 - x_0| + |x_0 - y| \leq 6|I| + \epsilon < 2\epsilon.$$

Since $|x - y|^{-1} < \epsilon^{-1}$, it follows that $E_2 < 2 f^*(z_0)$.

3.2 THEOREM. *If* $0 < p < \infty$, *then* $\|H^*f\|_p \leq C_p\|f^*\|_p$.

PROOF:

$$\|H^*f\|_p^p = p \int_0^\infty |\{H^*f > \lambda\}| \lambda^{p-1} d\lambda$$

$$= 2^p p \int_0^\infty |\{H^*f > 2\lambda\}| \lambda^{p-1} d\lambda$$

$$\leq 2^p p \int_0^\infty |\{H^*f > 2\lambda, f^* < \gamma\lambda\}| \lambda^{p-1} d\lambda + 2^p p \int_0^\infty |\{f^* > \gamma\lambda\}| \lambda^{p-1} d\lambda$$

$$\leq 2^p p C\gamma \int_0^\infty |\{H^*f > \lambda\}| \lambda^{p-1} d\lambda + \frac{2^p}{\gamma^p} \|f^*\|_p^p.$$

When we choose γ with $1 - 2^p C\gamma > 0$, we get that

$$\|H^*f\|_p \leq 2\gamma (1 - 2^p C\gamma)^{-\frac{1}{p}} \|f^*\|_p.$$

SECTION 3.2 THE A_p CONDITION

We now have control of H^*f on intervals, and we try to apply this to answer the question: for which $\omega > 0$ can we hope that

$$\int |Hf|^p \omega \leq C \int |f|^p \omega?$$

The trivial condition is that $a < \omega < b$, and since we can expect to control Hf on intervals, we could hope for a condition saying that ω should not deviate too much from its average on intervals; of course we hope that the deviation is controlled uniformly over all intervals. There are a lot of different ways of expressing this; we might hope to control $|\{x \in I | \omega(x) > \omega_I\}|$, or we could hope to bound $\int_I \omega; \int_I \omega^{-1}$ simultaneously.

3.3 DEFINITION: If ω is a locally integrable positive function, we say that $\omega \in A_p$ if:

when $p = 1$,

$$\omega_I \leq C \inf_{x \in I} \omega(x);$$

C independent of I.

when $1 < p < \infty$,

$$\left(\frac{1}{|I|}\int_I \omega\right)\left(\frac{1}{|I|}\int_I \omega^{\frac{1}{p-1}}\right)^{p-1} \leq C;$$

C independent of I.

when $p = \infty$,

$$\frac{\omega(E)}{\omega(I)} \leq \left(\frac{|E|}{|I|}\right)^\delta$$

for some δ independent of I and all measurable $E \subset I$.

3.4 REMARK: If $\omega \in A_\infty$, let

$$E = \{x \in I \mid \omega(x) > \beta\omega_I\}.$$

Then $|E| \leq \beta^{\frac{1}{1-\delta}}|I|$, and the A_∞ condition therefore implies that ω does not deviate from its average. Conversely, suppose that

$$\{x \in I \mid \omega(x) > \beta\omega_I\} \leq \beta^\delta|I|.$$

Then

$$\int_I \omega^{-\frac{1}{p-1}} = \frac{1}{p-1}\int \left|\{\omega^{-1} \geq \lambda\}\right| \lambda^{\frac{1}{p-1}-1}d\lambda$$

$$\leq \frac{1}{p-1}\int_0^{\omega_I^{-1}} |I|\lambda^{p-1}d\lambda$$

66

$$+p \int_{\omega_I^{-1}}^{\infty} \left(\lambda^{-1} \omega_I^{-1} \right)^{\delta} |I| \lambda^{p-1} d\lambda = C_{\delta} |I| \omega_I^{-\frac{1}{p-1}}.$$

Thus

$$\left(\frac{1}{|I|} \int_I \omega \right) \left(\int_I \omega^{-\frac{1}{p-1}} \right)^{p-1} \leq C_{\delta}^{p-1},$$

so that $\omega \in A_p$. For us, this computation means that the A_p condition is closely related to controlling the deviation of a function from its average.

3.5 LEMMA. If $\omega \in A_p$ then

$$|\{x \in I \mid \omega(x) \leq \beta \omega_I\}| \leq C \beta^{\frac{1}{p-1}} |I|.$$

PROOF: Rewriting the A_p condition,

$$C|I| \geq (\omega_I)^{\frac{1}{p-1}} \left(\int_I \omega^{-\frac{1}{p-1}} \right)$$

$$\geq (\omega_I)^{\frac{1}{p-1}} \left(\int_{\{\omega < \beta \omega_I\}} \omega^{-\frac{1}{p-1}} \right)$$

$$\geq (\omega_I)^{\frac{1}{p-1}} \left(\int_{\{\omega < \beta \omega_I\}} (\beta \omega_I)^{-\frac{1}{p-1}} \right)$$

$$= \beta^{-\frac{1}{p-1}} |\{x \in I \mid \omega < \beta \omega_I\}|.$$

3.6 REMARK: For functions which are essentially constant, there is no difference between L^p norms for different p's. The concept that A_p functions do not differ much from their averages allows us to prove basically the same result for A_p functions. We shall show that an A_p function is in A_∞ and also in $A_{p-\epsilon}$. Both facts follow from a precise statement of the equivalence of L^p norms. Remember that we average ω over intervals, so that the inclusion

$$L^p \subset L^{p+\epsilon}$$

follows from Holder's inequality. The suprising result that A_p is in $A_{p-\epsilon}$ follows from a "reverse Holders inequality".

3.7 THEOREM. Assume $\omega \in A_p$ for some p, $1 < p < \infty$. Then there exists a q, $q > 1$, with

$$\left(\frac{1}{|I|} \int_I \omega^q \right)^{\frac{1}{q}} \leq C \left(\frac{1}{|I|} \int_I \omega \right).$$

PROOF: We will first show that for some $\beta < 1$, we have a good-λ inequality:

$$\omega\{x \in I \mid \omega(x) > \lambda\} \leq C\lambda |\{x \in I \mid \omega(x) > \beta\lambda\}|.$$

Assuming this, the theorem is easy:

$$\int_I \omega^q = \beta^{-q} C_q \int_0^{\infty} \lambda^{q-1} |\{x \in I \mid \omega > \beta\lambda\}| \, d\lambda$$

$$\geq C'q \int_{\omega_I}^{\infty} \lambda^{q-2} \int_{\{x \in I \mid \omega \geq \lambda\}}^{\omega} (x) dx d\lambda$$

$$= C'q \int_{\{x \in I \mid \omega > \omega_I\}} \omega(x) \int_{\omega_I}^{\omega} \lambda^{q-2} d\lambda dx$$

$$= C'q \int_{\{x \in I \mid \omega > \omega_I\}} \omega(x) \left[\frac{\omega^{q-1}(x)}{q-1} - \frac{\omega_I^{q-1}}{q-1} \right] dx.$$

Now on the set

$$\{x \in I \mid \omega > \omega_I\},$$

$$-\omega \omega_I^{q-1} > \omega_I^q,$$

so that

$$C'q \int_{\{x \in I \mid \omega > \omega_I\}} \omega(x) \left[\frac{\omega^{q-1}(x)}{q-1} - \omega_I^{q-1}q - 1 \right] dx$$

$$\geq \frac{C'q}{q-1} \int_{\{x \in I \mid \omega > \omega_I\}} \omega^q - \omega_I^q.$$

But on $\{\omega \leq \omega_I\}$, the integrand is negative, and so this whole set may be added on without destroying the inequality:

$$\int_I \omega^q \geq \frac{C'q}{q-1} \int_I \omega^q - \omega_I^q,$$

whence,

$$\left(\frac{1}{|I|} \int_I \omega \right)^q \geq \left(\frac{C'q}{q-1} - 1 \right) \frac{1}{|I|} \int_I \omega^q.$$

The proof is completed if we take q close to 1.

There is also a good-λ inequality which analyzes the variation of ω over intervals. We control this through a Calderon-Zygmund decomposition: we construct the set $B = \{x \in I \mid \omega(x) > \lambda\}$, and let $\{J\}$ be disjoint intervals $J \subset I$ for which $\lambda \leq \omega_J \leq 2\lambda$. Then $\cup J = B$, and if $x \in I - B$, $\omega(x) \leq \lambda$. Let $\alpha = C\beta^{\frac{1}{p-1}}$, where C, β are as in Lemma 3.5. If $\beta < \beta_0$, then $\alpha < 1$, and

$$\omega\{x \in I \mid \omega > \lambda\} = \sum \omega(J) \leq 2\lambda \sum |J|.$$

But

$$|J| = |\{x \in J \mid \omega > \beta \omega_J\}| + |\{x \in J \mid \omega \leq \beta \omega_J\}|.$$

Since $\omega_J > \lambda$, this is dominated by

$$|\{x \in J \mid \omega > \beta \omega_J\}| + \alpha|J|.$$

It follows that

$$|J| \leq (1-\alpha)^{-1} |\{x \in J \mid \omega > \beta \lambda\}|,$$

and finally that

$$\omega\{x \in I \mid \omega > \lambda\} \le C\lambda \sum |\{x \in J \mid \omega > \beta\lambda\}|$$

$$= C\lambda |\{x \in B \mid \omega > \beta\lambda\}| \le C\lambda |\{xinI \mid \omega > \beta\lambda\}|.$$

3.8 COROLLARY. *If* $\omega \in A_p$, *then* $\omega \in A_\infty$.

PROOF: Choose q as in 3.7. Then:

$$\omega(E) = \int \chi_E \omega \le \|\chi_E\|_{q'} \|\chi_E \omega\|_q$$

$$\le |E|^{\frac{1}{q'}} |I|^{\frac{1}{q}} \left(\frac{1}{|I|} \int_I \omega^g \right)^{\frac{1}{q}}$$

$$\le C|E|^{\frac{1}{q'}} |I|^{\frac{1}{q}} \omega_I$$

$$= C|E|^{\frac{1}{q'}} |I|^{\frac{1}{q}-1} \int_I \omega = C \left(\frac{|E|}{|I|} \right)^{\frac{1}{q'}} \omega(I).$$

3.9 COROLLARY. *If* $\omega \in A_p$, *there is an* $\epsilon > 0$ *such that* $\omega \in A_{p-\epsilon}$.

PROOF: A change of notation shows that $w \in A_p$ if and only if

$$\omega^{-\frac{1}{p-1}} \in A_{p'}.$$

Applying 3.7 to $A_{p'}$, there is a $q > 1$ with

$$\left(\frac{1}{|I|} \int_I \omega^{-\frac{q}{p-1}} \right)^{\frac{1}{q}}$$

$$\le C \frac{1}{|I|} \int_I \omega^{-\frac{1}{p-1}} \le C \left(\frac{1}{|I|} \int_I \omega \right)^{-\frac{1}{p-1}},$$

since $\omega \in A_p$. In all, $\omega \in A_\gamma$ where $\frac{1}{\gamma-1} = \frac{q}{p-1}$; since $q > 1$, $\gamma < p$.

3.10 LEMMA. *If $\omega \in A_p$, then*

$$f_I \leq C \left(\frac{1}{\omega(I)} \int_I |f|^p \omega \right)^{\frac{1}{p}}.$$

If $\tilde{I}$ is the double of I (that is, $\tilde{I}$ has the same center as I but three times the length), then $\omega(\tilde{I}) \leq C\omega(I)$ and if J is the interval adjacent to I with the same length, then $\omega(J) \leq c\omega(I)$.

PROOF:

$$f_I = \frac{1}{|I|} \int_I f = \frac{1}{|I|} \int_I f \omega^{\frac{1}{p}} \omega^{-\frac{1}{p}}$$

$$\leq \frac{1}{|I|} \left(\int_I f^p \omega \right)^{\frac{1}{p}} \left(\int_I \omega^{-\frac{p'}{p}} \right)^{\frac{1}{p'}}.$$

But $\frac{p'}{p} = \frac{1}{p-1}$, and the second integral is dominated by

$$\left(\int_I \omega^{-\frac{p'}{p}} \right)^{\frac{1}{p'}} \leq C|I|^{\frac{1}{p'}} \left(\frac{\omega(I)}{|I|} \right)^{-\frac{1}{p(p-1)}}$$

$$= C|I|^{\frac{1}{p'}} |I|^{-\frac{1}{p}} [\omega(I)]^{-\frac{1}{p}}.$$

To get the results for $\tilde{I}$, let $f = \chi_I$. Then

$$\frac{1}{2} = f_{\tilde{I}} = C \left(\frac{1}{\omega(\tilde{I})} \int_{\tilde{I}} \chi_I \omega \right)^{\frac{1}{p}}$$

or,

$$C \leq \frac{\omega(I)}{\omega(\tilde{I})}.$$

A similar idea handles J.

3.11 LEMMA. *Let f be locally integrable and pick $\omega \in A_p$. Define*

$$f^{**}(x) = \sup_{x \in I} \frac{1}{\omega(I)} \int_I f(y)\omega(y) dy.$$

Then for $1 < p < \infty$,

$$\int |f^{**}|^p \omega \leq C \int |f|^p \omega.$$

PROOF: Recall the proof of Theorem 2.24; just as in that proof, we get a cover of $E = \{x | f^{**}(x) > \lambda\}$ by intervals, and we can apply the selection process to get disjoint intervals. We need that the measure of E is comparable to the sum of the measure of the disjoint intervals, but $E \subset \cup I^*$, so $\omega(E) \leq \sum \omega(I^*)$. But I^* has five times the length of I, so Lemma 3.10 tells us that

$$omega(E) \leq \sum \omega a(I^*) \leq C \sum \omega(I) \leq C \frac{1}{\lambda} \sum \int_I f,$$

and so on. In short, we can prove the maximal theorem 2.24 for measures besides Lebesgue measure, as long as they satisty the "doubling condition"

$$\omega(I^*) \leq C\omega(I).$$

70

3.12 THEOREM. *If $\omega \in A_p$ for $1 < p < \infty$, then*

$$\int |f^*|^p \omega \leq C \int |f|^p \omega.$$

PROOF: Since $\omega \in A_p$, there is a $\gamma < p$ with $\omega \in A_\gamma$. Then

$$f_I \leq C \left(\frac{1}{\omega(I)} \int_I f^\gamma \omega \right)^{\frac{1}{\gamma}},$$

whence

$$f^*(x) \leq C \left[(f^\gamma)^{**}(x) \right]^{\frac{1}{\gamma}}.$$

Thus,

$$\int |f^*|^p \omega \leq C \int |(f^\gamma)^{**}|^{\frac{p}{\gamma}} \omega.$$

Since $\frac{p}{\gamma} > 1$, 3.11 gives that the last integral is bounded by

$$C \int |f^\gamma|^{\frac{p}{\gamma}} \omega,$$

which proves the result.

3.13 THEOREM. *If $\omega \in A_p$, then*

$$\int |H^* f|^p \omega \leq C \int |f^*|^p \omega \leq C \int |f|^p \omega.$$

PROOF: Since $\omega \in A_p$, $\omega \in A_\infty$; choose δ as in the A_∞ condition. Apply a Whitney decomposition to $\{x | H^* f(x) > \lambda\}$, and let

$$E = \{x \in I \mid H^* f(x) > 2\lambda, \ f^*(x) < \gamma\lambda\}.$$

Then $|E| \leq C\gamma|I|$, from the good-λ inequality 3.1, and therefore

$$\omega(E) \leq \omega(I) \left[\frac{|E|}{|I|} \right]^\delta \leq C\gamma^\delta \omega(I).$$

Thus,

$$\omega\{x \in I \mid H^* f(x) > 2\lambda, \ f^*(x) < \gamma\lambda\}$$

$$\leq C\gamma^\delta \omega\{x \in I \mid H^* f(x) > \lambda\}.$$

Now the proof of L^p boundedness is exactly the same as in the proof of Theorem 3.2.

3.14 PROPOSITION. *If $1 < p < \infty$, and if*

$$\int |Hf|^p \omega \le C \int |f|^p \omega$$

for all $f \in L^p$, then $\omega \in A_p$.

PROOF: we let I, J be as in 3.10. Then if $x \in J$, $y \in I$, $|x - y| > 2|I|$, so that if f is supported in I,

$$|Hf(x)|\chi_J(x) = \left| \int \frac{f(y)}{x-y} dy \right| \chi_J(x) \ge (2|I|)^{-1} \int_I |f|\chi_J(x).$$

Then

$$\left(\int_J \omega\right)(|f_I|^p)$$

$$\le \int |Hf|^p \omega\chi_J \le \int |Hf|^p\omega$$

$$\le C \int |f|^p\omega = C \int_I |f|^p\omega.$$

In the special case $f = \chi_I$, $f_I = 1$,

$$\int_J \omega \le C \int_I \omega$$

so that, using the symmetry of I and J,

$$\omega(I) \le C\omega(J).$$

Then for arbitrary f supported in I,

$$\left(\int_I \omega\right)(|I|)^{-p}\left(\int_I \omega^{-\frac{1}{p-1}}\right)^p$$

$$\le C\left(\int_I \omega^{-\frac{p}{p-1}+1}\right),$$

etc, etc.

3.15 LEMMA. *If $\omega \in A_1, \omega \in A_p$.*

PROOF:

$$\left(\frac{1}{|I|}\int_I \omega\right)\left(\frac{1}{|I|}\int_I \omega^{-\frac{1}{p-1}}\right)^{p-1}$$

$$\le (\omega_I)\left(\frac{1}{|I|}\int_I 1\right)\left(\inf_I \omega\right)^{-1} = \frac{\omega_I}{\inf_I \omega} \le C.$$

3.16 COROLLARY. *Pick $s > 1$ and let $g \geq 0$ be a locally integrable function; define* $M_s(g) = [(g^s)^*]^{\frac{1}{s}}$. *Then $M_s(g) \in A_1$.*

PROOF: Let I and $\tilde{I}$ be fixed; let $\omega_1 = g^s \chi_{\tilde{I}}$ and $\omega_2 = \omega - \omega_1$. Since $\frac{1}{s} < 1$, we can use 2.27 to control ω_1^* :

$$\frac{1}{|I|} \int (\omega_1^*)^{\frac{1}{s}} \leq C \frac{1}{|I|} \int (\omega_1)^{\frac{1}{s}}$$

$$\leq C \frac{1}{|I|} \int_I (\omega_1)^{\frac{1}{s}} \leq C (\omega^*(y))^{\frac{1}{s}},$$

for any $y \in I$. Taking the infimum over all $y \in I$, we get

$$\left[(\omega_1^*)^{\frac{1}{s}} \right]_I \leq C \left[\inf_{y \in I} \omega^*(y) \right]^{\frac{1}{s}} = C \inf_{y \in I} [\omega^*]^{\frac{1}{s}}.$$

Thus ω_1 is controlled by an A_1 condition.

To control the ω_2 piece, choose x, $y \in I$ and compute that

$$\omega_2^*(x) = \sup_J \frac{1}{|J|} \int_J \omega_2.$$

If this is non-zero, the interval J must intersect the support of ω_2. On the other hand, J contains x, so J has length at least the length of I. Again, since $x \in J$, $\tilde{J}$ contains y, and therefore

$$\frac{1}{|J|} \int_J \omega_2 \leq 2 \frac{1}{|\tilde{J}|} \int_{\tilde{J}} \omega_2$$

$$\leq 2 \sup_J \frac{1}{|J|} \int_J \omega = 2\omega^*(y).$$

Now take the supremum over all J containing x to get $\omega_2^*(x) \leq 2\omega^*(y)$. Taking the infimum over all I containing y, we get

$$\omega_2^*(x) \leq 2 \inf_I \omega^*(y).$$

Thus

$$\frac{1}{|I|} \int_I (\omega_2^*)^{\frac{1}{s}} \leq 2 \inf_I (\omega^*)^{\frac{1}{s}} (y).$$

3.1): This result is due to R. Coifman. The proof here is taken from the paper of Coifman and C. Fefferman [7]; see the acknowledgements there. Again, results of this type extend to higher dimensions with similar proofs; as in the case of the maximal Hilbert transform, we have used pointwise estimates of K and the proofs do not apply to Hormander-type kernels.

3.2): The proof of L^p estimates for $0 < p < \infty$, using good-λ inequalities, is a technique from probability theory introduced by D. Burkholder, who gave probabilistic proofs of results related to these. See the survey article [2].

3.3): The A_p condition was discovered by B. Muckenhoupt [39] as part of a general study of the behaviour of maximal functions and Hilbert transforms on weighted L^p spaces. Special cases such as $|x|^\alpha$ were known for some time; see the discussion in Chapter IX of Hardy-Littlewood-Polya [26]. Our approach differs from Muckenhoupt's; again it is taken from [7]. For more problems concerning weighted norm inequalities see the survey article of Muckenhoupt [39], though a great deal of progress has been made on the problems mentioned there. The theory of A_2 weights plays an important role in the analysis of boundary values of harmonic functions; see the article of P. Jones [31] for a discussion.

3.16): This result was originally proved by A. Cordoba and C. Fefferman [14]; the proof given here is taken from R. Coifman and R. Rochberg [8].

CHAPTER 4 MULTIPLIERS WITH SINGULARITIES

In this Chapter we shall extend the standard model of Chapter 2. The model tells us that if $\mu = \sum \mu_j \phi_j$, where $\sum |\mu_j| < \infty$, then μ is a multiplier of all L^p. In the case of the Hilbert transform, $\sum |\mu_j| = \infty$; nonetheless the Hilbert transform is a multiplier of L^p for $1 < p < \infty$. The purpose of this chapter is to extend our understanding of the Hilbert transform to more general multipliers.

SECTION 4.1 THE HORMANDER-MIHILIN THEOREM

4.1 PROPOSITION. Assume $\mu \in L^\infty$ and $\mu' \in L^1$. Then μ is a multiplier of $L^p(\mathbb{R}^1)$ for $1 < p < \infty$.

PROOF: Let ψ be a smooth function with $\psi = 1$ on $[-N, N]$ and let $\psi = 0$ outside of $[-(N+1), (N+1)]$. Let $\mu_0 = \mu\psi$; we shall prove that $_p\|T_0\| \leq C$ where C is independent of N. This implies that μ is a multiplier by a standard limiting argument. As usual, we use duality tricks: let f, g be in $\mathcal{S}$, with $\|f\|_p = \|g\|_{p'} = 1$. Then

$$\int T_0 fg = \int \mu_0(\xi)\hat{f}(\xi)\hat{g}(\xi)d\xi = \int_{-\infty}^{\infty} \int_{-\infty}^{\xi} \mu_0'(\eta)d\eta \hat{f}(\xi)\hat{g}(\xi)d\xi$$

$$= \int_{-\infty}^{\infty} \int_{-\infty}^{\infty} \mu_0'(\eta)\chi_{(-\infty,\xi)}(\eta)\hat{f}(\xi)\hat{g}(\xi)d\xi d\eta$$

$$= \int_{-\infty}^{\infty} \int_{-\infty}^{\infty} \chi_{(\tau,\infty)}(\xi)\hat{f}(\xi)\hat{g}(\xi)d\xi \mu_0'(\eta)d\eta.$$

There are no boundary terms in the integration by parts because μ_0 vanishes at ∞; technical details of the interchange of integrals are easy since μ_0, μ_0' are in L^1 and $\hat{f}, \hat{g}$ are in $\mathcal{S}$. We now let T_η be the operator with multiplier

$$\chi_{(\tau,\infty)}(\xi) = \frac{1}{2}\left(1 - i\, isign(\xi - \eta)\right)$$

so that $T_\eta = \frac{1}{2}(I - iH_\eta)$, and H_η is a translate of the Hilbert transform. It follows that the T_η are uniformly bounded on L^p, and we can finish our estimates as

$$\int T_0 fg = \int_{-\infty}^{\infty} T_\eta f(x)g(x)dx\mu_0'(\eta)d\eta$$

$$\leq \int \|T_\eta\|_p \|g\|_{p'} |\mu_0'(\eta)|d\eta \leq C\|f\|_p \|g\|_{p'} \int |\mu_0'(\eta)|d\eta.$$

To complete the proof, note that

$$\|\mu_0'\|_1 \leq \|\mu'\psi\|_1 + \|\mu\psi'\|_1 \leq C\|\mu'\|_1 + \|\mu\|_\infty \|\psi'\|_1.$$

We have only to choose ψ with $\|\psi'\|_1 \leq C$, where C is independent of N. This is possible because ψ' is non-zero only on $(-(N+1), (N+1))$.

4.2 REMARK: We have changed our viewpoint a little; instead of looking at $\sum \mu_j \psi_j$ where the ψ_j are compactly supported, we are looking at weighted averages of Hilbert transforms. These two views are the same, just as an integral is a continuous version of a sum.

With this in mind, the condition that $\mu' \in L^1$ is analagous to the condition that $\sum |\mu_j| \leq \infty$. In short, the proposition hasn't really gotten us out of the standard model, other than in notation.

What we really want to know is the continuous analogue of the condition that $\mu = \sum \psi_j$ with no convergence factors out in front. This is not hard to see; intuitively, $\psi'_j = 2^j$ on the interval $[2^{-(j+1)}, 2^{-j}]$. In terms of μ, $\mu'(\xi) = \frac{1}{\xi}$. The correct multiplier condition is that $\xi \mu'(\xi) \in L^\infty$.

4.3 THEOREM. *Assume $\mu \in L^\infty$ and μ is C^1 away from 0. If $\xi \mu'(\xi) \in L^\infty$, then μ is a multiplier of L^p for $1 < p < \infty$.*

PROOF: We intend to apply Hilbert transform theory; we need to transfer our dyadic decomposition $\mu = \sum \psi_j$ over to estimates on convolution kernels. To do this, the decomposition of μ needs to be done smoothly, so that the inverse Fourier transform takes $\mathcal{S}$ into $\mathcal{S}$ functions. Technical details: let $\varphi \geq 0$ be smooth and let it satisfy

$$\varphi(\xi) = 1 \ if \ \frac{3}{4} < \xi < \frac{3}{2}$$

$$\varphi(\xi) = 0 \ if \ \xi < \frac{1}{2} \ or \ \xi > 2.$$

Let

$$\psi_j(\xi) = \frac{\varphi\left(2^{-j}|\xi|\right)}{\sum_{-\infty}^{\infty} \varphi\left(2^{-k}|\xi|\right)}.$$

Then let

$$\psi_j(\xi) = \psi_0\left(2^{-j}\xi\right);$$

let $\mu_j = \psi_j \mu$ and define the multiplier triples (T, K, μ) and (T_j, K_j, μ_j). Our hope is to apply Hilbert transform ideas, so we hope that

$$\int_{|x| > 2|y|} |K_j(x - y) - K_j(x)| \, dx \leq C 2^{-j}.$$

The uncertainty principle prevents this; since μ_j is supported on an interval of length 2^j, K_j is intuitively concentrated on an interval of length 2^{-j}. Therefore, if $|y| > 2^{-j}$, $K_j(x - y)$ cannot look at all like $K_j(x)$, so we cannot use control from smoothness. Our only hope is that K_j will satisfy the estimate:

$$\int_{|x| > 1} |K_j(x)| \, dx \leq C \left(\int_{|x| > 1} |x|^2 |K_j(x)| \, dx \right)^{\frac{1}{2}}$$

$$\leq C \|x K_j\|_2 = C \| \frac{d}{d\xi} \hat{K}_j \|_2 = C \| \mu'_j \|_2.$$

76

But
$$\| (\psi_j \mu)' \|_2 \leq \| \psi_j \mu' \|_2 + \| \psi_j' \mu \|_2$$
and
$$\int \left| D^{1-k} \psi_j D^k \mu \right|^2$$

$$= \int_{supp \ \psi_j} \left| 2^{-j(1-k)} \psi_0^{-k} \left(2^{-j} |\xi| \right) D^k \mu \right|^2$$

$$\leq C 2^{-2(1-k)j} \int_{supp \ \psi_j} \left| D^k \mu \right|$$

$$\leq C 2^{-2(1-k)j} \int_{2^{j-1}}^{2^{j+1}} |\xi|^{-2k} d\xi = C \left(2^{-\frac{j}{2}} \right)^2,$$

and therefore $\| x K_j \|_2 \leq C 2^{-\frac{j}{2}}$. We shall apply this type of estimate when no smoothness can be used; thus for $2^j |y| \geq 1$,

$$\int_{|x|>2|y|} |K_j(x-y) - K_j(x)| \, dx \leq$$

$$\int_{|x|>2|y|} |K_j(x-y)| \, dx + \int_{|x|>2|y|} |K_j(x)| \, dx.$$

In the first integral, a change of variables takes us to the set

$$\{x| \ |x+y| > 2|y|\} \subset \{|x| + |y| > 2|y|\} = \{|x| > |y|\},$$

and both integrals are dominated by

$$\int_{|x|>|y|} |K_j(x)| \, dx \leq \left(\int_{|x|>|y|} |x|^{-2} \right)^{\frac{1}{2}} \| x K_j \|_2$$

$$\leq C |y|^{-\frac{1}{2}} 2^{-\frac{j}{2}} = C \left(2^j |y| \right)^{-\frac{1}{2}}.$$

Next, in the region $|y| < 2^{-j}$, we expect to use smoothness. If $|y| < 2^{-j}$, but $|x| > 2|y|$, then

$$\int_{|x|>2|y|} |K_j(x-y) - K_j(x)| \, dx$$

$$\leq \left(\int_{|x| \leq 2^{-j}} 1 \right)^{\frac{1}{2}} \| K_j(x-y) - K_j(x) \|_2$$

$$= 2^{-\frac{j}{2}} \| \left(e^{2\pi i y \xi} - 1 \right) \psi_j(\xi) \mu(\xi) \|_2 \leq C 2^{-\frac{j}{2}} \| \xi \psi_j \mu \|_2.$$

On the support of ψ_j, $|\xi| \leq 2^j$ and $|\psi_j \mu| \leq C$ so this term is dominated by

$$C 2^{-\frac{j}{2}} 2^j |y| \, |support \ \psi_j|^{\frac{1}{2}} = C \left(2^j |y| \right).$$

Finally, if $|x| > 2^{-j}$,

$$\int_{|x|>2^{-j}} |K_j(x-y) - K_j(x)|\, dx$$

$$\leq \left(\int_{|x|>2^{-j}} |x|^{-2}\right)^{\frac{1}{2}} \|x\,[K_j(x-y) - K_j(x)]\|_2$$

$$= C2^{\frac{j}{2}} \| \left(e^{2\pi i y\xi} - 1\right) \left(\psi_j(\xi)\mu(\xi)\right)' \|_2 + 2^{-\frac{j}{2}} \| \left(e^{2\pi i y\xi} - 1\right) \psi_j(\xi)\mu(\xi)\|_2$$

$$\leq C2^{\frac{j}{2}}|y|\, \|\xi\,(\psi_j\mu)'\|_2 + C2^{\frac{j}{2}}|y|\, \|\psi_j\mu\|_2.$$

In the first term, $|\xi| \leq 2^j$ on *support* ψ_j, so that

$$\|(\psi_j\mu)'\|_2 \leq 2^{-\frac{j}{2}}.$$

In the second term,

$$\|\psi_j\mu\|_2 \leq \|\mu\|_\infty \|\psi_j\|_2 = C2^{\frac{j}{2}},$$

and both terms are dominated by $C2^j|y|$.

To finish the proof, let $\mu_N = \sum_{-N}^{N} \psi_j\mu$, and let (T_N, K_N, μ_N) be the corresponding multiplier triple. Since

$$\sum_{-N}^{N} \psi_j \leq 1, \quad \|\mu_N\|_\infty \leq \|\mu\|_\infty,$$

and we also have the main estimate

$$\int_{|x|>2|y|} |K_N(x-y) - K_N(x)|\, dx$$

$$\leq C \sum_{2^j|y|\leq 1} 2^j|y| + C \sum_{2^j|y|>1} \left(2^j|y|\right)^{-\frac{1}{2}}.$$

These sums are bounded independently of y; for example if $|y| > 1$, the exponents j in the first sum are all negative and contribute

$$C|y| \sum_{j\leq -\log|y|} 2^j = C|y||y|^{-1},$$

while in the second term,

$$|y|^{-\frac{1}{2}} \sum_{j\geq -\log|y|} 2^{-\frac{j}{2}}$$

$$= \left(\sum_{j=-\log|y|}^{\infty} 2^{-\frac{j}{2}} + \sum 2^{-\frac{j}{2}}\right) \left(|y|^{-\frac{1}{2}}\right) \leq C\left(1 + |y|^{-\frac{1}{2}}\right).$$

But $|y|^{-\frac{1}{2}} \leq 1$. A similar estimate works if $|y| \leq 1$.

Now we can just apply Theorem 2.32 to (T_N, K_N, μ_N), and get that they are bounded on L^p, with bounds independent of N. A quick limiting argument tells us how to extend the operator T to all of L^p. If $f \in \mathcal{S}$, $\sum \psi_j \leq 1$, and $\mu \in L^\infty$, so that $\mu_N \hat{f}$ converges dominatedly and also in L^2 to $\mu \hat{f}$. The operators T_N are uniformly bounded and converge to T on a dense set, hence T also is bounded. On $\mathcal{S}$, T is given by μ.

4.4 REMARKS: a) The Hilbert transform and its simple variant, given in 4.1, prove the boundedness of multipliers with simple jump discontinuities. Our new, improved result covers some oscillitory discontinuities; for example, $\mu(\xi) = |\xi|^{it}$ satisfies the hypotheses of the Theorem.

  b) Sometimes it is easier to check the weaker hypotheses that

$$\sup_{R>0} \frac{1}{R^{1-2j}} \int_{R<|\xi|<2R} |D^j \mu|^2 \leq C, \; j = 0, \; 1.$$

This was all we really needed in the proof of 4.4.

4.5 REMARK ON THE STANDARD MODEL

The proof we just did estimates $D^j \mu$ only on the support of f_j, which we took to be a dyadic interval. In this respect, we are viewing $\mu_j = \mu \psi_j$ as multipliers which behave independendly of each other. In the rest of this chapter, we are going to make that idea precise. As in the proofs above, we will do three versions. First we look at multipliers with no smoothness, like the Hilbert transform:

$$(P_j f)(\xi) = \chi_{(2^j, 2^{j+1})}(\xi) \hat{f}(\xi).$$

We'll prove a smooth version next, where $\varphi \in C^\infty$ and *support* $\varphi \subset [1, 2]$; then

$$(T_j f)(\xi) = \varphi(2^{-j}\xi) \hat{f}(\xi).$$

Finally, there will be a multiplier version:

$$(T_j f)(\xi) = \mu(\xi) \varphi(2^{-j}\xi) \hat{f}(\xi),$$

where μ obeys estimates similar to those in 4.3. As before, we will use the discontinuous P_j as a technical tool to prove multiplier theorems; this will finally eliminate kernel estimates on K and will allow us to concentrate totally on μ. The advantage in the end is that we can deal totally with the ideas in the standard model, and eliminate the need to switch back and forth between estimates on T and μ.

  The program is based on an understanding of what it means for the μ_j operators to act indepently of each other. Looking instead at the operators P_j with multiplier $\chi_{(2^j, 2^{j+1})}$, the independence is expressed by the fact that these operators are orthogonal projections:

$$\|f\|_2^2 = \|\hat{f}\|_2^2 = \sum_{-\infty}^{\infty} \int_{2^j}^{2^{j+1}} |\hat{f}|^2$$

79

$$= \sum \int |P_j f|^2 = \int \left(\sum |P_j f|^2 \right).$$

In terms of L^2 estimates, the correct expression of independence is:

$$\|f\|_2 = \| \left(\sum |P_j f|^2 \right)^{\frac{1}{2}} \|_2.$$

Now we have to guess the analogue of "independence" or "orthogonality" on L^p; our starting point is the quantity $(\sum |P_j f|^2)^{\frac{1}{2}}$, which is like $|P_j f|$ if not all the $|P_j f|$ are large at the same time. The theorems will give this as the correct idea of orthogonality:

$$\| \left(\sum |P_j f|^2 \right)^{\frac{1}{2}} \|_p \leq C_p \|f\|_p.$$

4.6 LEMMA. *Assume that*

$$\| \left(\sum |P_j f|^2 \right)^{\frac{1}{2}} \|_p \leq C_p \|f\|_p,$$

for all $f \in L^p$. Then

$$\|g\|_{p'} \leq C_p \| \left(\sum |P_j g|^2 \right)^{\frac{1}{2}} \|_{p'}$$

for all $g \in L^{p'}$.

PROOF:

$$\int fg = \int \hat{f}\hat{g} = \sum \int \widehat{P}_j f \widehat{P}_j g$$

$$= \int \sum (P_j f)(P_j g)$$

$$\leq \int \left(\sum |P_j f|^2 \right)^{\frac{1}{2}} \left(\sum |P_j g|^2 \right)^{\frac{1}{2}}$$

$$\leq \| \left(\sum |P_j f|^2 \right)^{\frac{1}{2}} \|_p \| \left(\sum |P_j g|^2 \right)^{\frac{1}{2}} \|_{p'}$$

$$\leq C_p \|f\|_p \| \left(\sum |P_j g|^2 \right)^{\frac{1}{2}} \|_{p'}.$$

Now we just take the supremum norm over all f with $\|f\|_p = 1$, and get the result.

4.7 REMARK: The whole point to this result is the corollary that, when we prove

$$\| \left(\sum |P_j f|^2 \right)^{\frac{1}{2}} \|_p \leq C_p \|f\|_p,$$

for $1 < p < \infty$, we automatically get

$$\|f\|_p \leq C_p \| \left(\sum |P_j f|^2 \right)^{\frac{1}{2}} \|_p \leq C_p \|f\|_p.$$

To compare the L^p norms of f and of $\sum |P_j f|^2$, we will need several new ideas; the complexity comes from the fact that we are defining an idea of orthogonality for Banach spaces L^p for $p \neq 2$, where orthogonality doesn't make any sense. The correct idea comes from probability theory, and is called independence. If f, g are measurable functions, we say that they are independent if $\int_E fg = \int_E f \int_E g$ for all measurable sets E; the idea is that the size of f has to influence the size of g pointwise, but on the average, over small sets, the functions do not influence each other. This is the type of object one would have to estimate if a sum to a non-integer power were expanded using the binomial theorem.

4.8 DEFINITION.

The Rademacher functions $r_m(t)$ are defined by the requirement that they be periodic of period 2, and that

$$r_0(t) = r_0(t + 1)$$

$$r_0(t) = 1 \ if \ 0 \le t \le \frac{1}{2}$$

$$r_0(t) = -1 \ if \ \frac{1}{2} < t \le 1$$

$$r_m(t) = r_0(2^m t).$$

4.9 THEOREM. The Rademacher functions are independent on $[0, 1]$; that is, for $m_1, m_2, \ldots, m_k, \ \alpha_1, \alpha_2, \ldots, \alpha_k$,

$$|\{t|r_{m_i}(t) \le \alpha_i \ for \ all \ i\}| = \prod_{i=1}^{k} |\{t|r_{m_i}(t) \le \alpha_i\}|,$$

and for measurable functions f_i and any measurable set E,

$$\int_E \prod f_i\left(r_{m_i}(t)\right) dt = \prod \int_E f_i\left(r_{m_i}(t)\right) dt.$$

PROOF: The result for integrals follows from a similar result for step functions, which follows from a result for Borel sets, which was born of a result for intervals, which was born of a result for half-infinite intervals, which we prove. We treat only the case $(-\infty, \alpha_i]$.

If $\alpha_i < -1$ for any i, some sets are empty on each side and the products are zero. If $\alpha_i > 1$ for any i, the condition $r_{m_i}(t) \le \alpha_i$ is always satisfied, so that it contributes nothing on the left-hand side and a factor of 1 on the right. Similarly, the condition $r_m(t) \le \alpha_m < 1$ is the same as $r_m(t) \le -1$ which is the same as $r_m(t) = -1$. The theorem will therefore follow from the equality

$$|\{t|r_{m_i}(t) = -1 \ for \ all \ i\}|$$

81

$$= \prod_{i=1}^{k} |\{t|r_{m_i}(t) = -1\}| = \left(\frac{1}{2}\right)^k.$$

A computation such as this is an induction, where all but the last part gets skipped. We reindex to get $m_1 < m_2 < \ldots < m_k$. The induction hypothesis is that $\{t|r_{m_i}(t) = -1, \ 1 \leq i \leq k-1\}$ is a disjoint union of dyadic intervals, each of which has length $2^{-m_{k-1}}$, and the total measure of the set is $\left(\frac{1}{2}\right)^{k-1}$. Now we add on the condition that $r_{m_k}(t) = -1$ on each of these dyadic intervals. Since $m_k > m_{k-1}$, r_{m_k} is negative half the time on each of these intervals(Proof? You must use the fact that r_{m_k} is a dilate of $r_{m_{k-1}}$).

4.10 THEOREM. Let $F(t) = \sum_{k=0}^{N} a_k r_k(y)$. If $\sum a_k^2 = 1$, then

$$|\{t \mid |F(t)| > \lambda\}| \leq 2e^{-\frac{\lambda^2}{4}}.$$

PROOF:

$$e^{\frac{\lambda^2}{2}} |\{t \mid |F(t)| > \lambda\}|$$

$$\leq \int_{|F|>\lambda} e^{\frac{\lambda}{2}|F(t)|} dt \leq \int_0^1 e^{\frac{\lambda}{2}|F|}$$

$$\leq \int_0^1 \left(e^{\frac{\lambda}{2}F} + e^{-\frac{\lambda}{2}F}\right) dt.$$

But

$$\int_0^1 e^{\frac{-\lambda}{2}F} = \int_0^1 \prod e^{-\frac{\lambda}{2}a_k r_k} dt$$

$$= \prod \int_0^1 e^{-\frac{\lambda}{2}a_k r_k} dt = \prod 2^{-k} \int_0^{2^{-k}} e^{-\frac{\lambda}{2}a_k r_0} dt$$

$$= \prod_{k=1}^{N} \left[\sum_{n=0}^{2^{k-1}} \left(\int_0^{n+\frac{1}{2}} e^{-\frac{\lambda}{2}a_k} dt + \int_{n+\frac{1}{2}}^{n+1} e^{\frac{\lambda}{2}a_k}\right)\right]$$

$$= \prod_{k=1}^{N} 2^{-k} \left(\frac{e^{\frac{\lambda}{2}a_k} + e^{-\frac{\lambda}{2}a_k}}{2}\right) \sum_{n=0}^{2^{k-1}} 1$$

$$= \prod_{k=1}^{N} \cosh(\frac{\lambda}{2}a_k) \leq \prod_{k=1}^{N} e^{\frac{\lambda^2}{4}a_k^2}$$

$$= e^{\frac{\lambda^2}{4}\sum a_k^2} = e^{\frac{\lambda^2}{4}}.$$

A similar estimate holds for the other term, which brings in the factor of 2.

4.11 COROLLARY. *Let $F(t) = \sum a_k r_k(t)$. If $1 < p < \infty$, then*

$$\|F\|_p \le C_p \|F\|_2 \le C_p' \|F\|_p.$$

PROOF: Let

$$G(t) = \|F\|_2^{-1} F(t) = \left(\sum a_k^2\right)^{-\frac{1}{2}} F(t).$$

Then $G = \sum b_k r_k$ where

$$\|G\|_2 = \left(\sum b_k^2\right)^{-\frac{1}{2}} = 1.$$

Therefore,

$$\|G\|_p^p = p \int_0^\infty \lambda^{p-1} |\{|G| > \lambda\}| \, d\lambda$$

$$\le 2p \int_0^\infty \lambda^{p-1} e^{-\frac{\lambda^2}{4}} \, d\lambda = C_p$$

and therefore $\|F\|_p \le C_p \|F\|_2$ for $1 \le p < \infty$. Repeating the proof of Lemma 4.6, we get $\|F\|_2^2 \le \|F\|_p \|F\|_r$ where $\frac{1}{2} = \frac{1}{2r} + \frac{1}{2p}$. If $p > 1$, $r < \infty$, and then $\|F\|_r \le C_p \|F\|_2$ for $1 \le p < \infty$.

REMARK: This, then, is the main point of independence: L^p norms are equivalent with L^2 norms. L^2 is a Hilbert space, where we can use orthogonality; the Corollary allows us to transfer that to L^p. The original proof expands the sum to a power using the binomial theorem. The integral of products simplify to the product of integrals, using independence. Normally, this can only be done on L^2, using orthogonality there; independence provides a substitute.

4.12 PROPOSITION. *Let $\{I_j\}$ be an arbitrary collection of intervals, and let*

$$(S_j f)(\xi) = \chi_{I_j}(\xi) \hat{f}(\xi).$$

Then, for $1 < p < \infty$,

$$\left\| \left(\sum |S_j f_j|^2\right)^{\frac{1}{2}} \right\|_p \le C_p \left\| \left(\sum |f_j|^2\right)^{\frac{1}{2}} \right\|_p.$$

PROOF: Since characteristic functins of intervals can be constructed from Hilbert transforms, we begin with a similar result for those. If $p > 2$, we let $r = \left(\frac{p}{2}\right)'$ and we choose any $g \in L^r$. Then for almost all x,

$$|g(x)| = \lim_{|I| \to 0} \left(\frac{1}{|I|} \int_I |g(y)|^s dy\right)^{\frac{1}{s}}$$

$$\le \sup_I \left(\frac{1}{|I|} \int_I |g(y)|^s dy\right)^{\frac{1}{s}} = M_s(g).$$

But Theorem 3.16 tells us that $M_s(g)$ is in A_1, and 3.15 tells us that $M_s(g)$ is in A_2, and 3.13 and 3.14 show that

$$\int |Hf|^2 M_s(g) \leq C \int |f|^2 M_s(g).$$

Therefore,

$$\int \left(\sum |Hf_j|^2\right) g \leq \int \left(\sum |Hf_j|^2\right) M_s(g)$$

$$\leq \left(\sum |f_j|^2\right) M_s(g) \leq C \|\sum |f_j|^2\|_{\frac{p}{2}} \|M_s(g)\|_{(\frac{p}{2})'}$$

$$= C \|\left(\sum |f_j|^2\right)^{\frac{1}{2}}\|_p^2 \|M_s(g)\|_r.$$

But $r > 1$, so we can find an $s < 1$ with $\frac{r}{s} > 1$ and 2.25 gives the estimate

$$\|M_s(g)\|_r^r = \int [(|g|^s)^*]^{\frac{r}{s}}$$

$$\leq C \int [|g|^s]^{\frac{r}{s}} = C \int |g|^r = C \|g\|_r^r.$$

In all,

$$\int \left(\sum |Hf_j|^2\right) g \leq C \|\left(\sum |f_j|^2\right)^{\frac{1}{2}}\|_p^2 \|g\|_r.$$

If we take the supremum over all g with $\|g\|_r = 1$, we obtain the $r' = \frac{p}{2}$ norm on the left-hand side, but this is just

$$\left(\int \left(\sum |Hf_j|^2\right)^{\frac{p}{2}}\right)^{\frac{2}{p}} = \|\left(\sum |Hf_j|^2\right)^{\frac{1}{2}}\|_p^2.$$

In the case of $p = 2$,

$$\|\left(\sum |Hf_j|^2\right)^{\frac{1}{2}}\|_2^2 = \int \sum |Hf_j|^2$$

$$= \int \sum |f_j|^2 = \|\left(\sum |f_j|^2\right)^{\frac{1}{2}}\|_2^2,$$

since h is an isometry on L^2. The case of $1 < p < 2$ is done using duality, just as in the proof of Lemma 4.6.

Next we need to go from H to intervals. Let $I_j = [a_j, b_j]$; then

$$S_j = \frac{1}{4}\left(I + iH_{a_j}\right)\left(I - iH_{-b_j}\right).$$

as in Lemma 2.11. Using the notation of Theorem 1.23(d), $H_c f = J(Hf)J^{-1}$ where $Jf = e^{2\pi cx} f$. It follows that

$$\left(I + iH_{a_j}\right)\left(I - iH_{-b_j}\right)$$

can be controlled by four operators, each of which is controlled by H or I.

4.13 THEOREM. Let $\Delta_j = (2^j, 2^{j+1}) \cup (-2^{j+1}, -2^j)$, for $-\infty < j < \infty$. Let P_j be the multiplier operator with multiplier χ_{Δ_j}. Then, for $1 < p < \infty$,

$$\|f\|_p \le C_p \| \left(\sum |P_j f|^2 \right)^{\frac{1}{2}} \|_p \le C_p' \|f\|_p.$$

PROOF: Exact equality holds at $p = 2$, so Lemma 4.6 tells us it is enough to prove the right-hand inequality. We shall use $\chi_{(-\infty,0)}$, which gives a bounded multiplier, to construct $(2^j, 2^{j+1})$ and its negative, separately. Our strategy is to use 4.12 to compare

$$\| \left(\sum |P_j f|^2 \right)^{\frac{1}{2}} \|_p$$

to

$$\| \left(\sum |T_j f|^2 \right)^{\frac{1}{2}} \|_p,$$

where the T_j are smooth. Then we will use 4.11 to compare

$$\| \left(\sum |T_j f|^2 \right)^{\frac{1}{2}} \|_p$$

to

$$\| \left(\sum T_j \right) f \|_p.$$

Finally, we will use 4.3 to control the smooth multiplier, $\sum T_j$, to bring us back to $\|f\|_p$. To further this nefarious scheme, we let $I_j = [2^j, 2^{j+1}]$ and we require $\psi \in S$ to satisfy

$$\phi(\xi) = 1 \; for \; 1 \le \xi \le 2$$

$$\phi(\xi) = 0 \; for \; \xi \; not \; in \; [\frac{1}{2}, \frac{5}{2}]$$

$$\phi_j(\xi) = \phi(2^j \xi).$$

Then $\chi_{I_j} = \psi_j \chi_{I_j}$, and if T_j denotes the operator with multiplier ψ_j, then $P_j f = P_j T_j f$, so that

$$\| \left(\sum |P_j f|^2 \right)^{\frac{1}{2}} \|_p = \| \left(\sum |P_j T_j f|^2 \right)^{\frac{1}{2}} \|_p$$

$$\le C \| \left(\sum |T_j f|^2 \right)^{\frac{1}{2}} \|_p.$$

Now let

$$\mu_o(\xi, t) = \sum_{j \; odd} \psi_j(\xi) r_j(t);$$

similarly for μ_e. We shall show that the μ satisfy the hypotheses of 4.3 uniformly in t. Let T_o, T_e be the corresponding operators. Then

$$\left(\sum |T_j f|^2 \right)^{\frac{1}{2}} \le C \| \sum T_j f(x) r_j(t) \|_{p,t}$$

by 4.11. Thus

$$\left(\sum |T_j f|^2\right)^{\frac{p}{2}} \leq C \int_0^1 \left|\sum T_j f(x) r_j(t)\right|^p dt$$

$$= C \int_0^1 |T_o f|^p dt \int \left(\sum |T_j f|^2\right)^{\frac{p}{2}} dx$$

$$\leq C \int_0^1 \int |T_o f(x)|^p dx dt \leq C \int_0^1 \int |f(x)|^p dx dt = C\|f\|_p^p.$$

This proves the theorem; to get the estimates on μ_o, note that the odd ψ_j have disjoint supports, so that

$$|\mu_o(\xi, t)| \leq \sum |\psi_j(\xi)| \leq 1.$$

Similarly, if $\xi \in \text{support } \psi_j$,

$$|\xi \mu_o'(\xi, t)| \leq |\xi \psi_j(\xi)| \, |\xi| \, 2^{-j} \, |\psi_0'(2^{-j}\xi)|$$

$$\leq 2^{j+1} 2^{-j} \|\psi_0'\|_\infty \leq C.$$

4.14 THEOREM. *Assume $\mu \in L^\infty$ and that for each dyadic interval I,*

$$\int_I |\mu'| \leq C$$

uniformly in I. Then μ yields an L^p multiplier for $1 < p < \infty$.

PROOF: First note that

$$\|Tf\|_p \leq C\| \left(\sum |P_j Tf|^2\right)^{\frac{1}{2}} \|_p.$$

If $I_j = \left(2^j, 2^{j+1}\right)$,

$$P_j Tf(x) = \int_{I_j} \mu(\xi)\hat{f}(\xi)e^{2\pi i x \xi} d\xi$$

$$= \int_{I_j} \int_{2^j}^\xi \mu'(\eta) d\eta \hat{f}(\xi)\chi_{I_j}(\xi)e^{2\pi i x \xi} d\xi + \mu(2^j)\int \hat{f}(\xi)\chi_{I_j}(\xi)e^{2\pi i x \xi} d\xi$$

$$= \int_{I_j} \mu'(\eta) \int_\eta^{2^{j+1}} \hat{f}(\xi)\chi_{I_j}(\xi)e^{2\pi i x \xi} d\xi + \mu(2^j)P_j f(x)$$

$$= \int_{I_j} \mu'(\eta)\frac{(I + iH_\eta)}{2} P_j f(x) d\eta + \mu(2^j)P_j f(x).$$

Since $\left|\mu(2^j)\right| \leq \|\mu\|_\infty$, we can isolate out the second term and then use the estimate

$$\| \left(\sum |P_j f|^2\right)^{\frac{1}{2}} \|_p \leq C\|f\|_p.$$

86

Since

$$\int_I \mu'(\eta) \frac{1}{2} I\left(P_j f(x)\right) d\eta = \frac{1}{2} \left(\int_I \mu'\right) P_j f(x) \le C P_j f(x),$$

this term may also be controlled. The "main" term is always the last one you can handle; in this case the main term in $|P_j T f|^2$ is

$$\left(\int_{I_j} \mu'(\eta) H_\eta P_j f(x) d\eta\right)^2$$

$$\le \left(\int_{I_j} |\mu'(\eta)|\right) \int_{I_j} |\mu'| \, |H_\eta P_j f(x)|^2 \, d\eta.$$

As in the proof of Proposition 4.12, we choose $g \in L^r$ for $r = \left(\frac{p}{2}\right)'$, and observe that $g \le M_s(g)$. Then

$$\int \sum |P_j T f|^2 g(x) dx$$

$$\le 2C \int \sum |P_j f|^2 g + C \sum \int_{I_j} |\mu'(\eta)| \int |H_\eta P_j f|^2 \, g(x) dx d\eta$$

$$\le C \| \left(\sum |P_j f|^2\right)^{\frac{1}{2}} \|_p^2 \, \|g\|_r$$

$$+ C \sum \int_{I_j} |\mu'(\eta)| \int |H_\eta P_j f|^2 \, M_s(g) dx d\eta.$$

The A_p theory applies to the H_η operators; the bounds are independent of η:

$$\int \sum |P_j T f|^2 g(x) dx \le C \|f\|_p^2 \|g\|_r + C \| \left(\sum |P_j f|^2\right)^{\frac{1}{2}} \|_p^2 \, \|M_s g\|_r.$$

Since we are proving a multiplier theorem, we may assume $p > 2$; there is an $s > 1$ with $\|M_s(g)\|_r \le C\|g\|_r$. In all,

$$\|Tf\|_p \le C \| \left(\sum |P_j f|^2\right)^{\frac{1}{2}} \|_p \le C\|f\|_p.$$

4.3): The results as stated and the proof given are taken from Hormander [28]; there is a similar result due to Mihilin [38]. The results are true in general dimension, although $|x|^{-1}$ is not integrable at infinity. This means that the general theorem needs more derivatives.

4.9): This result implies that the Rademacher functions are orthonormal, but they do not form a complete set.

4.11): The exposition here is taken from Stein [49]; for the original results see Paley[35].

4.12): The proof given here is modelled after results in A. Cordoba - C. Fefferman [14], where $M_s(g)$ was introduced. There is an alternative proof using the theory of vector-valued Hilbert transforms; see Stein [49].

4.13): This is called the Littlewood-Paley decomposition, after similar results for the circle. See the essay by Stein [48].

4.14): This is due to Marcinkiewicz [36]; again it is true in general dimension. It is important to note that there are no smoothness conditions at $\xi = 0$. The higher dimensional result has no smoothness conditions along co-ordinate axes. The proof here was shown us by Jodeit(unpublished).

CHAPTER 5 SINGULARITIES ALONG CURVES

In this chapter we continue our study of how singularities affect multipliers. We showed earlier that if μ is singular at a point or at dyadically spaced points, it can give rise to multipliers bounded on L^p. The theory extends immediately to $\mathbb{R}^2$; for example, $\mu(\xi_1, \xi_2) = isign\ (\xi_1)$ has a singularity along the line $\xi_1 = 0$, but it acts simply as a Hilbert transform in the y_1 variable:

$$Tf(x_1, x_2) = \lim_{\epsilon \to 0} \int_{|y_1| > \epsilon} f(x_1 - y_1, x_2) \frac{1}{y_1} dy_1.$$

Fubini's theorem therefore shows that T is a multiplier of $L^p(\mathbb{R}^2)$. Similarly, the characteristic function of the unit square is a multiplier, since it is a composition of rotated and translated one dimensional multipliers. The purpose of this chapter is to study singularities which are essentially two-dimensional; in particular, we shall study singularities along the curve $|\xi| = 1$.

SECTION 5.1 ASYMPTOTICS OF BESSEL FUNCTIONS

In this section we shall compute the Fourier transforms K_λ of the multipliers $\mu_\lambda(\xi) = \left(1 - |\xi|^2\right)_+^\lambda$; our goal is the estimate

$$K_\lambda(x) = K(x) + E(x);$$

here,

$$K(x) = C_\lambda |x|^{-\frac{n+1+2\lambda}{2}} \cos(2\pi |x| - (n+1)\pi - \frac{\lambda \pi}{2})$$

$$E(x) = O(|x|^{-\frac{n+3+2\lambda}{2}}).$$

Following the treatment in Stein and Weiss, we give a detailed derivation of these estimates from properties of Bessel functions.

5.0 FACT.

$$J_\lambda(R) = C_\lambda R^\lambda \int_{-1}^1 e^{iRt} \left(1 - t^2\right)^{\lambda - \frac{1}{2}} dt.$$

PROOF: This was discussed in Proposition 2.3

5.1 LEMMA. If $d\theta$ denotes the surface measure on the unit sphere S^{n-1}, then

$$d\breve{\theta}(x) = \int_{S^{n-1}} e^{2\pi i x \xi} d\theta(\xi)$$

$$= c_n \frac{J_{\frac{n-2}{2}}(2\pi |x|)}{|x|^{\frac{n-2}{2}}}.$$

PROOF: We let $x = |x|x'$ where $|x'| = 1$. We parametrize the sphere by spheres of one lower dimension; in $\mathbb{R}^3$ this would be by circles parallel to the equator, but of varying radius. If the north pole is fixed as $\bar{n}$, then the vector $\bar{\xi}'$ with tip on the unit sphere and lying along one of these circles is characterized by the condition $\bar{n} \cdot \bar{\xi}' = constant$. The poles of course have $\bar{n} \cdot \bar{\xi}' = 0$; the equator $\bar{n} \cdot \bar{\xi}' = 1$.

The north pole can be chosen arbitrarily; to make life simple we think of $x \in \mathbb{R}^n$ as fixed, and we arrange our north pole to be $\bar{x}'$. Then for $\bar{\xi}'$ with tip on the sphere, the angle θ between $\bar{x}'$ and ξ' is in $[-\pi, \pi]$. If we let $L_\theta = \{\bar{\xi}' \in S^{n-1} | \bar{x}' \cdot \bar{\xi}' = \cos\theta\}$, then this gives a decomposition of S^{n-1} into S^{n-2} spheres. These are spherical-polar coordinates with $\bar{x}'$ as the north pole. The sphere L_θ has radius $|\sin\theta|$, so L_θ has volume $c_n |\sin\theta|^{n-2}$. Finally, we can compute:

$$d\breve{\theta}(x) = \int_{-\pi}^{\pi} \int_{L_\theta} e^{2\pi i |x| \bar{x}' \cdot \bar{\xi}'} d\xi d\eta$$

$$= \int_{-\pi}^{\pi} \int_{L_\theta} e^{2\pi i |x| \bar{x}' \cdot \bar{\xi}'} d\eta d\theta.$$

But for $\bar{\xi}' \in L_\theta$, $\bar{x}' \cdot \bar{\xi}' = \cos\theta$ and this is independent of η, so

$$d\breve{\theta}(x) = \int_{-\pi}^{\pi} e^{2\pi i |x| \cos\theta} \int_{L_\theta} 1 d\eta d\theta$$

$$= \int_{-\pi}^{\pi} e^{2\pi i |x| \cos\theta} |L_\theta| d\theta = c_n \int_{-\pi}^{\pi} e^{2\pi i |x| \cos\theta} |\sin\theta|^{n-2} d\theta$$

$$= c_n \int_{-1}^{1} e^{2\pi i |x| y} |1 - y^2|^{\frac{n-3}{2}} dy$$

$$= c_n \frac{J_{\frac{n-2}{2}}(2\pi |x|)}{|x|^{\frac{n-2}{2}}}.$$

5.2 PROPOSITION.

$$\breve{\mu}_\lambda(x) = K_\lambda(x)$$

$$= c_\lambda \frac{J_{\frac{n}{2}+\lambda}(2\pi |x|)}{|x|^{\frac{n}{2}+\lambda}}.$$

PROOF: Apply power series techniques to Bessel's differential equation to obtain

$$J_\mu(t) = \sum_{0}^{\infty} \frac{(-1)^j (\frac{t}{2})^{\mu+j}}{\Gamma(j+\mu+1)},$$

whence

$$\int_0^1 J_\mu(st) s^{\mu+1} (1-s^2)^\nu ds$$

$$= \sum_{0}^{\infty} \frac{(-1)^j (\frac{t}{2})^{\mu+j}}{\Gamma(j+\mu+1)} \int_0^1 s^{\mu+2j} s^\mu (1-s^2)^\nu s ds$$

90

$$= \sum_{0}^{\infty} \frac{(-1)^j (\frac{t}{2})^{\mu+j}}{\Gamma(j+\mu+1)} \frac{1}{2} \int_0^1 r^{\mu+j}(1-r)^\nu s \, ds.$$

Recalling the Beta integral, this is

$$\frac{1}{2} \sum_{0}^{\infty} \frac{(-1)^j (\frac{t}{2})^{\mu+j}}{j!} \frac{\Gamma(\nu+1)\Gamma(\mu+j+1)}{\Gamma(\mu+\nu+1+j+1)\Gamma(\mu+j+1)}$$

$$= \frac{\Gamma(\nu+1)2^{\nu+1}}{t^{\nu+1}} \sum_{0}^{\infty} \frac{(-1)^j (\frac{t}{2})^{\mu+\nu+1+j}}{\Gamma(j+1+\nu+\mu+1)}$$

$$= c_\nu \frac{J_{\mu+\nu+1}(t)}{t^{\nu+1}}.$$

In our case, we are evaluating $K_\lambda(x)$, which is a Fourier transform. Using polar co-ordinates and the Bessel function expression for $d\check{\theta}(x)$, we get

$$K_\lambda(x) = c_\lambda \int_0^1 (1-r^2)^\lambda r^{n-1} J_{\frac{n-2}{2}}(2\pi r|x|) (r|x|)^{\frac{n-2}{2}} \, dr$$

$$= c_\lambda |x|^{\frac{n-2}{2}} (2\pi|x|)^{-(\lambda+1)} J_{\frac{n-2}{2}+\lambda+1}(2\pi|x|).$$

5.3 PROPOSITION.

$$J_\lambda(r) = c_\lambda r^{-\frac{1}{2}} \cos\left(r - \frac{\pi\lambda}{2} - \frac{\pi}{4}\right) + O\left(r^{-\frac{3}{2}}\right).$$

PROOF: The standard trick in this subject is to use the analyticity of the integrand in the expression

$$r^{-\lambda} J_\lambda(r) = c \int_{-1}^1 e^{irs}(1-s^2)^{\lambda-\frac{1}{2}} ds$$

to change contours; if this is done cleverly, the term $(1-s^2)^{\lambda-\frac{1}{2}}$ becomes like $s^{\lambda-\frac{1}{2}}$. This is much better, because we can use homogeneity to evaluate the dependence on r. Of course the phase of the exponential will change with the contour, and this accounts for the cos terms in the expansion.

To work, then. We write

$$\int_{-1}^1 e^{irs}(1-s^2)^{\lambda-\frac{1}{2}} ds = \lim_{\epsilon \to 0} \int_{-1+\epsilon}^{1-\epsilon} e^{irs}(1-s^2)^{\lambda-\frac{1}{2}} ds,$$

which allows us to avoid the points of non-analyticity. Then $e^{irz}(1-z^2)^{\lambda-\frac{1}{2}}$ has an analytic branch on the simply connected domain $\mathbb{C} - \{(-\infty, -1+\epsilon] \cup [1-\epsilon, \infty)\}$. We are going to be a bit sloppy with the ϵ's in the following.

We take a rectangle with height a and base $[-1, 1]$, and traverse the boundary counterclockwise. The boundary curve γ has four pieces:

$$\gamma_1 = \{s + i0| -1 \le s \le 1\}; \quad \gamma_2 = \{1 + iy| 0 \le y \le a\};$$

91

$$\gamma_3 = \{s + ia| -1 \le s \le 1\}; \quad \gamma_4 = \{-1 + i(a - y)| \ 0 \le y \le a\}.$$

Then

$$\int_{\gamma_1} = -\int_{\gamma_2} - \int_{\gamma_3} - \int_{\gamma_4},$$

and we take the limit as $a \to \infty$. Then

$$\int_{\gamma_3} = -\int_{-1}^{1} e^{-ar} e^{irs} \left(1 - (s + ia)^2\right)^{\lambda - \frac{1}{2}} ds,$$

and the exponential guarentees convergence to zero. For γ_2,

$$\left(1 - (i + iy)^2\right)^{\lambda - \frac{1}{2}} = \left(1 - (1 + y)^2\right)^{\lambda - \frac{1}{2}}$$

$$= \left(y^2 - 2iy\right)^{\lambda - \frac{1}{2}} = y^{\lambda - \frac{1}{2}}(y - 2i)^{\lambda - \frac{1}{2}},$$

and similarly for γ_3. Together, these two integrals contribute

$$\int_0^\infty e^{ir(1+y)} y^{\lambda - \frac{1}{2}}(y - 2i)^{\lambda - \frac{1}{2}} dy - \int_0^\infty e^{ir(-1+y)} y^{\lambda - \frac{1}{2}}(y + 2i)^{\lambda - \frac{1}{2}} dy.$$

We now expand $(y + 2i)^{\lambda - \frac{1}{2}}$ into a power series, with first term $(2i)^{\lambda - \frac{1}{2}}$; in each integrand this contributes a term like $2^{\lambda - \frac{1}{2}} e^{(\lambda - \frac{1}{2})\frac{i\pi}{2}}$; together the two integrals produce

$$\cos\left(r - \frac{\pi\lambda}{2} - \frac{\pi}{4}\right) \int_0^\infty e^{ry} y^{\frac{1}{2} + \lambda} dy$$

$$= \Gamma(\lambda + \frac{1}{2}) r^{-(\frac{1}{2} + \lambda)} \cos\left(r - \frac{\pi\lambda}{2} - \frac{\pi}{4}\right).$$

The proof will be done if we can control the terms arising from the errors, $(y + 2i)^{\lambda - \frac{1}{2}} - (2i)^{\lambda - \frac{1}{2}}$. If $y \le 1$, we use smoothness, and the mean value theorem bounds the difference by Cy. If $y > 1$, we factor out the y and again use the mean value theorem. Then

$$\int_0^\infty E(y) e^{-ry} y^{\lambda - \frac{1}{2}} dy \le \int_0^1 e^{-ry} y^{\lambda + \frac{1}{2}} dy + \int_1^\infty e^{-ry} y^{2\lambda - 1} dy$$

$$= r^{-(\lambda + \frac{3}{2})} \int_0^r e^{-y} y^{-(\lambda + \frac{1}{2})} dy + r^{-2\lambda} \int_r^\infty e^{-y} y^{2\lambda + 1} dy$$

$$= r^{-(\lambda + \frac{3}{2})} \Gamma(\lambda + \frac{3}{2}) + C_k r^{-2\lambda} \int_r^\infty y^{-k} y^{2\lambda + 1} dy = O\left(r^{-(\lambda + \frac{3}{2})}\right).$$

5.4 COROLLARY. T_λ is bounded on $L^p(\mathbb{R}^n)$ for $1 \le p \le \infty$ if $\lambda > \frac{n-1}{2}$, and, T_λ is not bounded on $L^p(\mathbb{R}^n)$ for $\lambda < \frac{n-1}{2}$ if p is not in the range

$$\frac{2n}{n+1+2\lambda} < p < \frac{2n}{n-1-2\lambda}.$$

(cf Figure 2).

PROOF: If $\lambda > \frac{n-1}{2}$, $K_\lambda \in L^1$. To get the unboundedness results, we would like to take f to be a delta function; the asymptotics would show that

$$K_\lambda * f \approx |x|^{-(\frac{n+1+2\lambda}{2})},$$

and this cannot be in L^p unless $\frac{2n}{n+1+2\lambda} < p$. Since delta functions aren't in L^p, we have to be more clever; the key is to take f of such small support that the oscillation in the convolution kernels can be supressed. If we think of f as compactly supported, the convolution averages f against translates of K_λ, so there will be times when the average is almost zero. We have to analyze the places where f comes together with the peaks of K_λ.

Let D be a ball of radius $\frac{1}{100}$, and let A_k denote the annulus

$$\left\{ x \;\middle|\; \left| 2\pi|x| - \lambda\pi - \frac{\pi}{2} - 2k\pi \right| \le 10^{-2} \right\}.$$

Note that the measure of A_k is about Ck^{n-1}, and that if $x \in A_k$, $y \in D$, then

$$\cos\left(2\pi|x-y| - (n+1)\pi - \frac{1}{2}(\frac{n}{2}+\lambda)\pi \right) \ge \frac{1}{2}.$$

Then, if we take $f = \chi_D$,

$$\|T_\lambda f\|_p \ge \|(K_\lambda * f)\chi_{\cup A_k}\|_p - \|(E * f)\chi_{\cup A_k}\|_p.$$

But

$$(K_\lambda * f)(x)\chi_{\cup A_k}(x) = \int_D \frac{\cos(2\pi|x-y| + mess)}{|x-y|^{\frac{n+1+2\lambda}{2}}} dy$$

$$\ge Ck^{-(\frac{n+1+2\lambda}{2})};$$

similarly,

$$|E * f|\chi_{\cup A_k} \le Ck^{-(\frac{n+3+2\lambda}{2})}.$$

Finally,

$$\|T_\lambda f\|_p \ge C \left(\sum k^{-\frac{n+1+2\lambda}{2}} |A_k| \right)^{\frac{1}{p}}.$$

This sum is finite if and only if $(n-1) - (n+1+2\lambda)\frac{p}{2} < -1$.

5.5 REMARK: We already see that singularities along curves are much worse than those along lines. The characteristic function of the unit square has convolution kernel

$$\frac{\sin x_1}{x_1} \frac{\sin x_2}{x_2}$$

and this is in L^p for all $p > 1$. In contrast, the characteristic function of the unit disc has transform

$$\frac{J_1(|x|)}{|x|}$$

and this is in L^p only when $p > \frac{4}{3}$. We have seen detailed proofs of both these results; the geometry underlying the distinction eludes us.

93

Our standard way to access L^p is through interpolation from L^1 and L^2; since T_λ is unbounded on L^1 for small lambda, this approach fails. We fall back on the standard model: we extend our smooth compactly supported functions ψ_j to $\mathbb{R}^n$ by starting with ψ_0 defined on $\mathbb{R}^1$ and then letting $\psi_j(\xi) = \psi_0\left(2^j(|\xi| - 1)\right)$. The ψ_j have support in an annulus of approximate thickness 2^{-j}, centered at $|\xi| = 1$. (This is rather different from the one dimensional picture, but remember that the singularity we hope to resolve is at $|\xi| = 1$.) We expect then that $\mu_\lambda \approx 2^{-j\lambda}$ and that $_p\|T_\lambda\| \leq \sum 2^{-j\lambda} {}_p\|T_j\|$. Unlike the one-dimensional case, however, the ψ_j are not dilates of a single function, and the T_j therefore do not have equal operator norms. To get control of these norms, we look at norms on L^1 and L^2 and interpolate.

On L^2, $_2\|T_j\| = \|\psi_j\|_\infty = 1$. On L^1, $_1\|T_j\| = \|\check\psi_j\|_1$. But intuitively, $\check\psi_j$ is just the Fourier transform of the characteristic function of the disc, restricted to the annulus $\{2^j < |x| < 2^{j+1}\}$. A quick computation shows that we expect $\|\check\psi_j\|_1 = C 2^{\frac{n-1}{2}j}$. The computation can also be done more precisely, using asymptotics:

$$\check\psi_j(x) = \int \frac{J_{\frac{n-2}{2}}(2\pi|x|r)}{(|x|r)^{\frac{n-2}{2}}} \psi\left(2^j(|r| - 1)\right) r^{n-1} dr.$$

In the region $|x| < 1$, $|\check\psi_j| \leq \|\psi_j\|_1 \leq 1$; in the region $|x| > 1$ and for those $r \in support\ \psi_j$, r is close to 1, and therefor $r|x| > 1$, so that we are in the region in which the asymptotics for Bessel functions hold. The main term contributes

$$C|x|^{-\frac{n-2}{2}} \int e^{2\pi i r|x|} \psi\left(2^j(|r| - 1)\right) r^{\frac{n-1}{2}} dr.$$

Let $s = 2^j(|r| - 1)$; we get

$$C|x|^{-\frac{n-2}{2}} e^{2\pi i |x|} 2^{-j} \int e^{2\pi i s 2^{-j}|x|} \psi(s)(1 + s2^{-j})^{\frac{n-1}{2}} ds.$$

If we expand $(1 + s2^{-j})^{\frac{n-1}{2}}$ about $s = 0$, the main term is

$$C|x|^{-\frac{n-2}{2}} e^{2\pi i |x|} 2^{-j} \int e^{2\pi i s 2^{-j}|x|} \psi(s) ds$$

$$= C|x|^{-\frac{n-2}{2}} e^{2\pi i |x|} 2^{-j} \check\psi_0(2^{-j}|x|).$$

Notice that $\check\psi_0 \in \mathcal{S}$, and, all in all, if we let

$$\varphi(x) = |x|^{-\frac{n-1}{2}} \check\psi_0(|x|),$$

then $\varphi \in L^1$, and

$$|\check\psi_j(x)| \leq C 2^{\frac{n-1}{2}j} 2^{-jn} \varphi(2^{-j}x).$$

94

Then $\|\tilde{\psi}_j\|_1 \leq C2^{\frac{n-1}{2}j}\|\varphi\|_1 = C2^{\frac{n-1}{2}j}$. This affirms our intuition, though the reader should be warned that we have been very sloppy controlling the error terms in the above computation.

We can now interpolate: if $\frac{1}{p} = \frac{t}{1} + \frac{1-t}{2}$, or, $t = \frac{2}{p} - 1$, so

$$_p\|T_j\| \leq {}_1\|T_j\|^t {}_2\|T_j\|^{1-t}$$
$$\leq 2^{j(\frac{n-1}{2})t}1^t \leq 2^{j(n-1)(\frac{1}{p}-\frac{1}{2})}.$$

This allows us to sum the geometric series if

$$\lambda > (n-1)\left(\frac{1}{p} - \frac{1}{2}\right),$$

if $p \leq 2$.

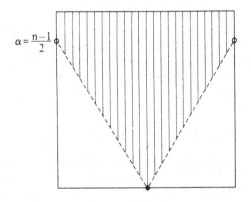

$$\alpha = \frac{n-1}{2}$$

Figure 3

5.6 THEOREM. *If* $0 < \lambda < \frac{n-1}{2}$, *then* T_λ *is bounded on* $L^p(\mathbb{R}^n)$ *for* $\lambda > (n-1)\left|\frac{1}{p} - \frac{1}{2}\right|$ *(cf Figure 3).*

PROOF: Of course we can give a proof along the outlines of the standard model; just to show a variety of techniques, we'll give a proof using a decomposition of the K_λ instead of μ_λ.

The standard decomposition suggests that we decompose K_λ into pieces supported on dyadic annuli. To this end, let $K \in \mathcal{S}$ satisfy $0 \leq K \leq 1, K(x) = 1$ for $|x| \leq 100; K(x) = 0$ for $|x| > 200$. Let $\phi_j(x) = K\left(\frac{|x|}{2^j}\right) - K\left(\frac{|x|}{2^{j+1}}\right)$. Notice that ϕ_j is supported on the annulus $100 \cdot 2^j < |x| < 200 \cdot 2^j$, so that ϕ_j acts like a smoothed version of the characteristic function of an annulus.

We now let $K_j = \phi_j K_\lambda$. Then $\sum \phi_j(x) = K(x) - K(0)$, so that $K(x) - \sum \phi_j(x) = K(0) = 1$, and therefore $K_\lambda = K_\lambda - \sum K_j$. Since $K_\lambda - K$ is bounded and compactly supported, it yields an L^1 error term, which is a bounded operator on all L^p.

In this picture, it is easy to compute $\|K_j\|_1$, and hard to do $\|\hat{K}_j\|_\infty$. The first gives us:

$$\int |K_\lambda(x)| \, |\phi_j(x)| \, dx \leq C \int_{100 \cdot 2^j}^{200 \cdot 2^j} r^{-\frac{n+1+2\lambda}{2}} r^{n-1} dr = C2^{-\lambda j}2^{j\frac{n-1}{2}},$$

95

as before. To get an intuitive sense of how large $\|\hat{K}_j\|_\infty$ might be, note that $\hat{K}_j = \hat{K}_\lambda \star \hat{\phi}_j$, but

$$\hat{\phi}_j(\xi) = 2^{nj}\hat{K}(2^j\xi) - 2^{n(j-1)}\hat{K}(2^{j-1}\xi).$$

If we write $\delta = 2^{-j}$, $\psi(\xi) = 2\hat{K}(\xi) - \hat{K}(\tfrac{\xi}{2})$, then we can rewrite things as

$$\psi_\delta(\xi) = \delta^{-n}\psi(\delta^{-1}\xi) = \hat{\phi}_j(\xi).$$

Now ψ_δ is a standard approximate identity, except for the fact that $\int \psi_\delta = 0$. As δ gets small, *i.e.* as $j \to \infty$, we expect that $\psi_\delta \star \mu_\lambda$ will converge to $\left(\int \psi_\delta\right)\mu_\lambda = 0$, and we expect that the rate of convergence will be controlled by the smoothness of μ_λ. Thus, we expect the estimate:

$$|\psi_\delta \star \mu_\lambda - 0| \le C\delta^\lambda = C2^{-j\lambda}.$$

There is a more intuitive way to think of these things; think of ψ as being the characteristic function of a disc; then $\psi_\delta \star f$ averages f over a smaller disc. Near $|\xi| = 1$, the largest value of μ_λ is $C\delta^\lambda$, and this dominates the average. Away from $|\xi| = 1$, μ_λ is smooth, so that the average is like $\mu_\lambda \int \psi_\delta = 0$.

This completes the intuitive analysis; as in the argument at the beginning of the section, the estimates:

$$\|K_j\|_1 \le C2^{j\frac{n-1}{2}}2^{-j\lambda}; \quad \|\hat{K}_j\|_\infty \le C2^{-j\lambda}$$

are enough to finish the proof of the theorem using interpolation. Our duty now is to give a rigouous, albeit tedious, interpretation of the intuition behind these estimates. Several important tricks come out, though, so the proof is a "must-see".

In the region $|\xi| > 2$, the "supports" of ψ_δ and μ_λ are disjoint, so $\psi_\delta \star \mu_\lambda$ should be zero. Since ψ_δ is not actually compactly supported, we need another trick. The rapid decrease of $\psi \in \mathcal{S}$ should do it. Thus,

$$|\psi_\delta(\eta)| = |\delta^{-n}\psi(\delta^{-1}\eta)|$$

$$\le C\delta^{-n}|\delta^{-1}\eta|^{-(n+i)} |K_j(\xi)| \le \int |\mu_\lambda(\xi - \eta)|\,|\psi_\delta(\eta)|\,d\eta$$

$$\le C\|\mu_\lambda\|_\infty \delta^{-i} \int \chi_D(\xi - \eta)|\eta|^{-(n+i)}d\eta;$$

here $D = support\ \mu_\lambda$. Since we must have $|\xi - \eta| \le 1$, $|\xi| - |\eta| < 1$ or $|\eta| > |\xi| - 1 > 2 - 1$, and the integral is bounded by $C\delta^{-i}$, which is small if we choose i large enough.

In the region $|\xi| < 2$, we expect to control the size of $\psi_\delta \star \mu_\lambda$ by the rate of convergence of $\psi_\delta \star \mu_\lambda$ to 0. typically, we would estimate

$$\left| \psi_\delta \star \mu_\lambda(\xi) - \left(\int \psi_\delta\right)\mu_\lambda(\xi) \right|$$

$$\le \int |\mu_\lambda(\xi - \delta\eta) - \mu_\lambda(\xi)|\,|\psi(\eta)|\,d\eta.$$

96

Now we want to use the estimate

$$|\mu_\lambda(\xi - \delta\eta) - \mu_\lambda(\xi)| \le C|\delta\eta|^\lambda$$

to finish the proof. Unfortunately, this estimate is wrong when $\lambda > 1$; however, if $\lambda \le 1$ we can finish the proof:

$$\int |\mu_\lambda(\xi - \delta\eta) - \mu_\lambda(\xi)|\, |\psi(\eta)|\, d\eta$$

$$\le C \int |\delta\eta|^\lambda |\psi(\eta)| d\eta \le C\delta^\lambda.$$

If $\lambda > 1$, we need to integrate by parts $k = [\lambda]$ times; this will differentiate μ_λ into a mess times $(1 - |\xi|^2)^{\lambda-k}$, and now $\lambda - k \le 1$. For this to work, the kth integral of ψ, call it ϕ, must have the same structure as ψ, that is, $\phi \in \mathcal{S}$ and $\int \phi = 0$. Certainly this won't hold for the indefinite integral of an arbitrary ψ; we need to be careful. Let ϕ be defined by

$$\check{\phi}(x) = \frac{\check{\psi}(x)}{(2\pi i x_1)\ldots(2\pi i x_k)}.$$

Now $\check{\psi} = K_1$, which vanishes in a neighborhood of the origin, so that $\check{\phi} \in \mathcal{S}$, $\phi \in \mathcal{S}$; $\int \phi = \check{\phi}(0) = 0$. Moreover, $D_{\xi_1} \ldots D_{\xi_k}\phi = \psi$, and

$$\int \eta_1 \ldots \eta_k \phi(\eta) d\eta = \left(D_{x_1} \ldots D_{x_k}\check{\phi}\right)(0) = 0.$$

Moreover,

$$D_{\eta_1} \ldots D_{\eta_k}\mu_\lambda(\xi - \delta\eta) = C(\delta\eta_1)\ldots(\delta\eta_k)\mu_{\lambda-k}(\xi - \delta\eta).$$

Thus we may integrate by parts k times and generate no boundary terms, since $\mu_{\lambda-k}$ and ϕ both vanish at infinity. We want to note that these formulae are not accurate if the η's are not distinct; on the other hand we only want to do the case $\lambda < \frac{n-1}{2} < n$, so we have room to magesterially choose the η's distinct. To finish the proof:

$$\left|\hat{K}_j(\xi)\right| = \left|\int \mu_\lambda(\xi - \delta\eta)\psi(\eta)d\eta\right|$$

$$= C\delta^k \left|\int \mu_{\lambda-k}(\xi - \delta\eta)\eta_1 \ldots \eta_k\phi(\eta)d\eta - \mu_{\lambda-k}\int \eta_1 \ldots \eta_k\phi(\eta)d\eta\right|$$

$$\le C\delta^k \int |\mu_{\lambda-k}(\eta - \delta\eta) - \mu_{\lambda-k}(\eta)|\,|\eta_1 \ldots \eta_k|\,|\phi(\eta)|d\eta$$

$$\le C\delta^k\delta^{\lambda-k} \int |\eta|^{\lambda-k}|\eta|^k|\phi(\eta)|d\eta \le C\delta^\lambda.$$

A quick look at the straight lines in Figure 3 suggests that an interpolation has happened. But what kind of interpolation? After all, Figure 3 does not graph the L^p, L^q boundedness of some operator, but instead the L^p, L^p boundedness of a family of operators. To interpolate, one would need some convexity of the $_p\|T_\lambda\|$ as a function of λ. If we study the proof of the Riesz-Thorin theorem, the requirement is that $F(z) = \int T_z fg$ should be analytic on a strip, continuous on the boundary, and grow no faster than $Ce^{-b|y|}$. We call such families of operators admissible.

97

5.7 THEOREM. *Let T_z be an admissible family of operators in the region $0 \le Re z \le 1$. Assume that T_{0+iy} is bounded on L^{p_0}, T_{1+iy} on L^{p_1}. If*

$$_{p_j}\|T_{j+iy}\| \le e^{Ce^{-b|y|}}$$

for some $b < \pi$, then T_t is bounded on L^p for $\frac{1}{p} = \frac{t}{p_0} + \frac{1-t}{p_1}$.

PROOF:

This result, called complex interpolation, is due to Stein. Although this is an important result, detailed proofs would take us too far afield; see the references at the end of the chapter.

To prove the theorem, Stein first extends the three-lines theorem to obtain the estimate

$$\log|F(z)| \le \frac{1}{2}\sin\pi x \int_{-\infty}^{\infty} \frac{\log|F(iy)|}{\cosh\pi y - \cos\pi x} + \frac{\log|F(1+iy)|}{\cosh\pi y + \cos\pi x} dy.$$

Stein now takes $f = \sum a_j \chi_{E_j}$; $g = \sum b_k \chi_{F_k}$ to be simple functions with $\|f\|_p = \|g\|_{p'} = 1$. Using the notation of Theorem 1.13, we let

$$F(z) = \int T_z \left(\sum |a_j|^{p\alpha} e^{i\theta_j} \chi_{E_j} \right) \left(\sum |b_k|^{p'(1-\alpha)} e^{i\psi_k} \chi_{F_k} \right)$$

where $\alpha(z) = \frac{1-z}{p_0} + \frac{z}{p_1}$. Then

$$|F(1+iy)| \le {}_{p_1}\|T_{1+iy}\| \le M_1(y)$$

$$|F(0+iy)| \le {}_{p_0}\|T_{0+iy}\| \le M_0(y)$$

as in the proof of 1.13; here the $M_j(y)$ do not grow more quickly than $e^{Ce^{-b|y|}}$. The modified three-lines theorem therefore gives us

$$|F(z)| \le \exp \frac{1}{2}\sin\pi x \int_{-\infty}^{\infty} \frac{\log|M_0(y)|}{\cosh\pi y - \cos\pi t}$$

$$+ \frac{\log|M_1(y)|}{\cosh\pi y + \cos\pi t} dy$$

and the growth of $M_j(y)$ guarantees the convergence of the integral. Thus $|F(z)| \le C_t$ as desired.

5.8 COROLLARY. *Assume $0 < \lambda < \frac{n-1}{2}$. Then T_λ is bounded on L^p for $\lambda > (n-1)\left|\frac{1}{p} - \frac{1}{2}\right|$.*

PROOF: Let

$$K_z(x) = c_z \frac{J_{\frac{n}{2}+z}(2\pi|x|)}{|x|^{\frac{n}{2}+z}}$$

where c_z remains to be specified. To check admissibility,

$$\int T_z(f)g \le \|T_z f\|_2 \|g\|_2 \le \|f\|_2 \|g\|_2.$$

Similarly,
$$2\|T_{0+iy}\| = \|\mu_{0+iy}\|_\infty \leq 1.$$

But
$$1\|T_{1+iy}\| \leq \|K_{\lambda+iy}\|_1.$$

Now the normalization constants become important! From Lemma 5.1, 5.2, the standard normalizations for Bessel functions involve quotients of gamma functions. If one looks at the definition of the gamma function as an integral, one sees that it grows exponentially in the complex plane. A similar growth may be seen for the Bessel function itself; one need only look at the power series expansion. Thus it is important to balance these two growths; see Stein [45] for the actual work.

5.4): The simple counter-example here only works because the oscillation

$$\cos(2\pi|x|)$$

of the kernel is slow; it would not have worked on the superficially similar kernels with oscillation $\cos(2\pi|x|^2)$. The L^p behaviour of such kernels is very different; see C. Fefferman [22] and S. Wainger [53] for a treatment of $\frac{e^{i|x|^\alpha}}{|x|^\beta}$.

5.5): We take up the local computations of K_λ again in Chapter 6 Section 1.

5.6): The proof given here is due to deLeeuw [19].

5.7): This result is due to Stein [45]; for other techniques of interpolation, see the discussion in Stein and Weiss [50], p. 209.

5.8): This result is due to Stein [45].

CHAPTER 6 RESTRICTION THEOREMS

The purpose of this chapter is to prove the L^p boundedness of the Bochner-Riesz operators T_λ in the optimal range $\frac{2n}{n+1+2\lambda} \leq p \leq \frac{2n}{n-1-2\lambda}$, but only if λ is fairly large. The problem is in controlling the oscillation: in one dimension, we replaced the convolution kernel $\frac{\sin 2\pi x}{x}$ by $\frac{1}{x}$, but in higher dimensions the exponential $e^{i|x|}$ cannot be simplified. We shall describe the theory of C. Fefferman and E. M. Stein which analyzes oscillatory kernels of this type.

SECTION 6.1 RESTRICTION

To derive a simple model of the effects of oscillation, we analyze a simple kernel in $\mathbb{R}^2$:

$$K(x) = \frac{e^{i|x|}}{|x|^{\frac{3}{2}+\lambda}}.$$

Assume that f is compactly supported in $D = \{y|\ |y| < 1\}$, and assume that x is large. The geometry which controls $|x - y|$ is illustrated in Figure 4. We may take $|x - y|$ to be approximately equal to $|x|$ minus the length of the projection of y onto x. Thus,

$$|x - y| \approx |x| - y \cdot \frac{x}{|x|}$$

$$e^{i|x-y|} \approx e^{i|x|} e^{-iy \cdot \frac{x}{|x|}}.$$

We can also estimate

$$|x - y|^{\frac{3}{2}+\lambda} \approx |x|^{\frac{3}{2}+\lambda} + errors,$$

and then

$$K \star f(x) = \int \frac{e^{i|x-y|}}{|x-y|^{\frac{3}{2}+\lambda}} f(y) dy \approx \frac{e^{i|x|}}{|x|^{\frac{3}{2}+\lambda}} \int e^{-iy \cdot \frac{x}{|x|}} f(y) dy$$

$$= \frac{e^{i|x|}}{x|^{\frac{3}{2}+\lambda}} \hat{f}\left(\frac{x}{|x|}\right).$$

The counterexample of Corollary 5.4 led us to expect that $T_\lambda f$ would look like $|x|^{-(\frac{3}{2}+\lambda)}$, but this new estimate is even nicer: it separates out the action of K into two components: a Fourier transform in the spherical variable, and a purely radial decrease at infinity. In polar co-ordinates, we expect then that

$$\|K \star f\|_p^p \geq \int_2^\infty \int_0^{2\pi} |K \star f(r, \theta)|^p\, d\theta r\, dr$$

$$= \int_2^\infty r r^{-(\frac{3}{2}+\lambda)p} dr \approx \int_0^{2\pi} \left|\hat{f}(1, \theta)\right|^p d\theta.$$

It follows that $K \star f$ is not in L^p unless $p > \frac{4}{3+2\lambda}$; if this is satisfied, then the L^p boundedness of K would imply that

$$\int_0^{2\pi} \left|\hat{f}(1,\theta)\right|^p d\theta \leq C\|f\|_p^p.$$

An inequality of this type is called a restriction theorem, because it governs the restriction of the Fourier transform of f to the set $\{\xi \mid |\xi| = 1\}$, which is, after all, a set of measure zero. This is remarkable, for an arbitrary measurable function cannot be restricted to sets of measure zero without infinities. On the other hand, if $f \in L^p$ for $1 < p \leq 2$, $\hat{f}$ is computed from a limiting process in $L^{p'}$ norm: if $f_n \in \mathcal{S}$, and $f_n \to f$, then the Hausdorff-Young inequality

$$\|\hat{f}\|_{p'} \leq \|f\|_p$$

guarantees that the $\hat{f}_n$ converge in $L^{p'}$-norm to some function $\hat{f}$. A subsequence converges almost everywhere, so that $\hat{f}$ is defined almost everywhere. But defining $\hat{f}$ on a set of measure zero seems ridiculous.

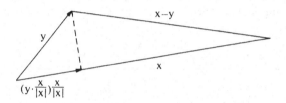

Figure 4

Since the restriction theorems are suprising enough to cast suspicion on their validity, and raise doubts on the boundedness of the T_λ, we start with restriction theorems.

6.1 PROPOSITION. *Let f be a radial function on $\mathbb{R}^n$. Then*

$$\left|\hat{f}(1,\theta)\right| \leq C\|f\|_p$$

for all radial f in L^p , if and only if $1 \leq p < \frac{2n}{n+1}$.

PROOF: Since f is radial, $\hat{f}$ is radial, and $\hat{f}(1,\theta)$ is independent of θ. Then the linear functional $f \to \hat{f}(1,\theta)$ is bounded on L^p if and only if it is given by integration against an $L^{p'}$ function. We abbreviate $\hat{f}(1,\theta)$ by $\hat{f}(1)$; then

$$\hat{f}(1) = \int_0^\infty f(r) \int_{S^{n-1}} e^{2\pi\sigma r\xi} d\sigma r^{n-1} dr$$

$$= \int_0^\infty f(r) \frac{J_{\frac{n-2}{2}}(2\pi r|\xi|)}{(r|\xi|)^{\frac{n-2}{2}}} r^{n-1} dr.$$

But

$$\frac{J_{\frac{n-2}{2}}(2\pi r)}{(r)^{\frac{n-2}{2}}}$$

is in $L^{p'}$ if and only if $p' > \frac{2n}{n-1}$. Thus the result.

This is only a first approximation to a theorem, since it treats only radial functions. Nonetheless, it shows that even these have Fourier transforms with suprising smoothness properties(up to a point). Our next result gives an attempt at a geometric understanding of the critical index $\frac{2n}{n+1}$ that appears in restriction; we want to limit the range of (p, q) for which restriction from L^p to L^q can hold. The first estimates we had on the Fourier transform were in the Hausdorff-Young theorem; we showed that these were optimal using the dilation structure of the Fourier transform: $f(\delta x)$ was transformed into $\delta^{-1}\hat{f}(\delta^{-1}\xi)$. But circles are not dilation invariant, and we must be sneakier.

Intuitively we want to make $\hat{f}|_{S^1}$ as large as possible. Utilizing dilation, we imagine that $\hat{f}$ is supported on a small rectangle which we orient to intersect as much of S^1 as possible. See Figure 5. What dimensions on the support of $\hat{f}$ maximize $\hat{f}|_{S^1}$ while minimizing $\|f\|_p$? Take the rectangle to be of size $\delta \times \delta^\alpha$, where $\alpha < 1$, so that this is the longer side. We choose α so that the corner of the rectangle just intersects the circle; this maximizes intersection with the circle. Since $x = \sqrt{1-y^2} = \sqrt{1-(1-\delta)^2} \approx \sqrt{\delta}$, we take $\alpha = \frac{1}{2}$.

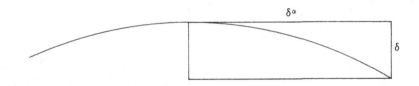

Figure 5

6.3 PROPOSITION. *The inequality*

$$\|\hat{f}|_{S^{n-1}}\|_q \le C\|f\|_p$$

is false if $q > \frac{n-1}{n+1}p'$.

PROOF: Let ψ be a smooth compactly supported function with $\psi(\xi) = 1$ if $|\xi| < \sqrt{n}$. Define

$$\hat{f}_\delta(\xi_1,\ldots,\xi_n) = \psi\left(\delta^{-2}(\xi_1 - 1), \delta^{-1}\xi_2,\ldots,\delta^{-1}\xi_n\right).$$

If $\delta < 1, \xi \in S^{n-1}$, $|\xi_j| \le \frac{\delta}{\sqrt{n}}$, we claim that $\hat{f}_\delta = 1$. It is enough to show that $\delta^{-2}|\xi_1 - 1| < 1$, or, $1 - \xi_1 < \delta^2$. But

$$1 - \xi_1 = 1 - \sqrt{1 - (\xi_2^2 + \ldots + \xi_n^2)} \le 1 - \left(1 - (\xi_2^2 + \ldots + \xi_n^2)\right)$$

103

$$= \xi_2^2 + \ldots + \xi_n^2 \leq (n-1)\left(\frac{\delta}{\sqrt{n}}\right)^2 \leq \delta^2.$$

Therefore,

$$\|\hat{f}|_{S^{n-1}}\|_q \geq \left|\{(\xi_1,\ldots,\xi_n)| \,|\xi_j| \leq \frac{\delta}{\sqrt{n}}\}\right|^{\frac{1}{q}}$$

$$\geq C\left(\frac{\delta}{\sqrt{n}}\right)^{\frac{n-1}{q}} = C\delta^{\frac{n-1}{q}}.$$

On the other hand,

$$f_\delta(x_1,\ldots,x_n) = \delta^{n+1}e^{2\pi i\delta^2}x_1\check{\psi}(\delta^2 x_1, \delta x_2, \ldots, \delta x_n)$$

so that

$$\|f_\delta\|_p = \delta^{\frac{n+1}{p'}}\|\check{\psi}\|_p = C\delta^{\frac{n+1}{p'}}.$$

Thus, for all small δ, $\delta^{\frac{n+1}{p'}} \geq C\delta^{\frac{n-1}{q}}$, whence the theorem.

6.4 REMARK: In Figure 6 we sketch the region in which restriction is known to fail. The dotted line represents the boundary of the region in the above result; the vertical line comes from the requirement $1 \leq p < \frac{2n}{n+1}$, which is necessary even for radial functions. The two lines intersect at $p = q = \frac{2n}{n+1}$.

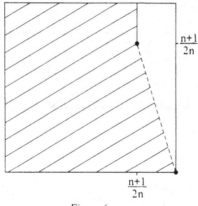

$\frac{n+1}{2n}$

$\frac{n+1}{2n}$

Figure 6

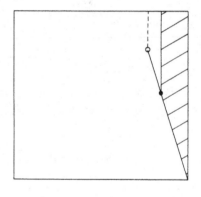

Figure 7

6.5 PROPOSITION. If $1 \leq p \leq \frac{2(n+1)}{n+3}$, then

$$\|\hat{f}|_{S^{n-1}}\|_2 \leq C\|f\|_p.$$

PROOF: We may assume that $f \in \mathcal{S}$, and this is justification for the formal manipulations below. Then

$$\|\hat{f}|_{S^{n-1}}\|_2^2 = \int \hat{f}(\sigma)\hat{f}(\sigma)d\sigma$$

$$= \int (f \star f)\widehat{d\sigma} = \int f \star f(x)\widehat{d\sigma}(x)dx$$

104

$$= \int \int f(x-y)f(y)dy\widehat{d\sigma}(x)dx$$

$$= \int \widehat{d\sigma} \star f(x)f(x)dx$$

$$\leq \|f\|_p \, \|\widehat{d\sigma} \star f\|_{p'}.$$

It suffices to prove that $Rf = \widehat{d\sigma} \star f$ is a bounded map from L^p to $L^{p'}$. We have two techniques for doing this; one is to chop $\widehat{d\sigma}(x)$ into pieces; the other is to use complex interpolation. Here we do the interpolation; in the discussion following the theorem we see what the standard model has to tell us.

We need to embed the operator R into an analytic family of operators; we let $R_z f = K_z \star f$; as usual we can only control what happens at $p = 1$ and $p = 2$. Then at $p' = \infty$; $p' = 2$ we use

$$\|K_z \star f\|_\infty \leq \|K_z\|_\infty \, \|f\|_1$$

$$\|K_z \star f\|_2 \leq \|\hat{K}_z\|_\infty \, \|f\|_2.$$

Ah, it's our old enemy, simultaneous control of K and of $\hat{K}$! Now,

$$K_0 = \widehat{d\sigma}(x) = \frac{J_{\frac{n-2}{2}}(2\pi|x|)}{|x|^{\frac{n-2}{2}}},$$

and not only is this in L^∞, but $|x|^{\frac{n-1}{2}}K_0 \in L^\infty$. So, we can afford to make K_0 a lot worse and still keep the first inequality. On the other side, $\hat{K}_0 = d\sigma$, and this is not an L^∞ function at all, so it must be improved considerably. Between the improvement and the worsening will fall K_0, and we will get a range of p for which K_0 gives a bounded operator. Let

$$K_z(x) = c_z \frac{J_{\frac{n-2}{2}+z}(2\pi r|\xi|)}{(r|\xi|)^{\frac{n-2}{2}+z}}.$$

If $z = -\frac{n-1}{2}$, $\frac{n-2}{2} + z = -\frac{1}{2}$ and we note from formula 3.4(6) of Watson that

$$J_{-\frac{1}{2}}(2\pi|x|) = \frac{\cos 2\pi|x|}{\pi|x|^{\frac{1}{2}}}.$$

Thus, at $z = -\frac{n-1}{2}$, K_z is in L^∞ (indeed, this is what we would expect from asymptotics for Bessel functions).

On the other hand, if $z > 0$, $\hat{K}_z(\xi) = (1-|\xi|^2)_+^{z-1}$, and this is in L^∞ for the first time when $z - 1 = 0$. Complex interpolation gives $z = \frac{n+3}{2} - \frac{n+1}{p}$. At $z = 0$, this gives the theorem.

REMARKS: a)In Figure 7, we sketch the region of boundedness for restriction. Open circles and dotted lines represent regions of known unboundedness.

b) If f is replaced by $\delta^{-n}f(\delta^{-1}x)$, $\hat{f}$ is replaced by $\hat{f}(\delta\xi)$, and $\hat{f}_\delta|_{S^{n-1}}$ is the same as $\hat{f}$ restricted to a sphere of radius δ^{-1}. Following through the changes of variables gives us

$$\left(\int_{S^{n-1}}\left|\hat{f}(R,\sigma)\right|^2 d\sigma\right)^{\frac{1}{2}} \leq C_p R^{-\frac{n}{p'}}\|f\|_p.$$

6.7 ESSAY ON RESTRICTION: A weaker version of restriction can be proved using the standard model; as in the proof of Theorem 5.6, we write $\widehat{d\sigma} = \sum K_j$. Since

$$|\widehat{d\sigma}(x)| \leq C|x|^{-\frac{n-1}{2}},$$

$$\|K_j\|_\infty \leq C2^{-j\frac{n-1}{2}}.$$

To estimate $\|\hat{K}_j\|_\infty$, we rely on the ideas in 5.6: $\hat{K}_j = \psi_\delta \star d\sigma$, where $\delta = 2^{-j}$. Intuitively ψ_δ is compactly supported on a ball of radius δ, so that $\psi_\delta \star d\sigma$ is zero outside of the annulus $1 - \delta < |\xi| < 1 + \delta$. To maximize $\psi_\delta \star d\sigma$, we look at $|\xi| = 1$; then $\psi_\delta \star d\sigma$ should be viewed as an average of ψ_δ against $d\sigma$. That is, it is the average of a ball of radius δ over the surface of a sphere. The ball intersects the sphere in a region of surface measure δ^{n-1}; on the other hand the ball itself has measure δ^n, so that the average is $\delta^{-1} = 2^j$. We therefore expect:

$$\|K_j\|_\infty \leq C2^{-j\frac{n-1}{2}}$$

$$\|\hat{K}_j\|_\infty \leq C2^j.$$

Then the norm of the operator given by convolution with K_j is $2^{\epsilon j}$ where $\epsilon = 2\left(1 - \frac{1}{p}\right) - \frac{n-1}{2}\left(1 - 2(1 - \frac{1}{p})\right)$. This is negative when $p < 2\frac{n+1}{n+3}$.

Explaining restriction by our standard model still hasn't told us what it all means. We will weave a fairly complicated argument suggesting some of the ideas which make restriction work. Still, today no one can give a geometric criterion for deciding which sets of measure zero enjoy restriction and which do not, so we can hardly say we understand restriction.

We will first examine the effect of curvature. If restriction is valid for some curve $S \subset \mathbb{R}^2$, and S contains a line segment, we can assume by translation, rotation, dilation and chopping off, that $S = [0,1] \times \{0\}$. Now let $f(x_1, x_2) = g(x_1)h(x_2)$, and note that $\hat{f}(\xi_1, \xi_2)|_S = \hat{g}(\xi_1)|_I\hat{h}(0)$, so that if we choose $g \in S$ fixed with $\hat{g}|_I = 1$, then a restriction inequality would imply that $|\hat{h}(0)| \leq C\|h\|_p$. Thus the map $h \to \hat{h}(0)$ is a bounded linear functional on L^p, which implies that $e^{2\pi i 0 x}$ is in $L^{p'}$. This only works if $p = 1$. The moral is that restriction cannot happen for any surface which contains a piece of a hyperplane.

Our second observation relates restriction to S and the Fourier transform of the measure supported on S. We let S be a smooth compact surface and let $d\sigma$ be the surface measure; we claim that the inequality $\|\hat{f}|_S\|_q \leq C\|f\|_p$ is valid for some $p > 1$ if and only if $\widehat{d\sigma} \in L^{p_0}$ for some $p_0 < \infty$.

106

First assume restriction holds. Define $R : L^p(\mathbb{R}^n) \to L^q(S)$ by $Rf = \hat{f}|_S$. Then the adjoint operator $R^* : L^{q'}(S) \to L^{p'}(\mathbb{R}^n)$ is also bounded, and we quickly compute that

$$\int (R^* g)(x)f(x)dx = \int g(\sigma)Rf(\sigma)d\sigma$$

$$= \int g(\sigma)\hat{f}(\sigma)d\sigma = \int [\widehat{g(\sigma)d\sigma}](x)f(x)dx.$$

Thus, $R^* g = \widehat{gd\sigma}$; this has to be interptreted in the sense of distributions, and, for our purposes, we may as well assume $g \in C^\infty(S)$. Choosing $g \equiv 1$,

$$\|\widehat{d\sigma}\|_{p'} \leq C|S|^{\frac{1}{q'}}.$$

If $p > 1, p' < \infty$.

Conversely, assume that $\widehat{d\sigma} \in L^{p_0}$, for some $p_0 < \infty$. Then

$$\int |\hat{f}|^2 d\sigma = \int f * \hat{f} d\sigma$$

$$\leq \|\widehat{d\sigma}\|_{p_0} \|f * f\|_{p_0'} \leq C\|f\|_p^2;$$

here, we have used Young's convolution inequality

$$\|h * g\|_r \leq \|h\|_p \|g\|_q$$

$$1 + \frac{1}{r} = \frac{1}{p} + \frac{1}{q}.$$

Since $r = p_0'$, $p = q$, $p = \frac{2p_0}{2p_0-1} > 1$.

Since the L^p behaviour of the Fourier transform of $d\sigma$ captures the restrictability of S, a geometric understanding of restriction would have to begin with how the curvature of S affects the size of $\widehat{d\sigma}$. The correct result is that $\widehat{d\sigma} \in L^{p_0}$ if the curvature of S does not vanish to infinite order at any point of S. Rather than attempt to prove this here, we will see how the standard model treats the difference between a line and a circle. Let ϕ be a smooth function with $\phi(x) = 1$ in the range $|x| < 1$, $\phi(x) = 0$ for $|x| > 2$. Then as $\delta \to 0$, $\delta^{-1}\phi(\delta^{-1}x_2)\phi(x_1)$ converges in distribution to the measure supported on $[-1,1] \times \{0\}$. The Fourier transforms act like

$$\hat{\phi}(\xi_1)\hat{\phi}(\delta\xi_2) \to \hat{\phi}(\xi_1)\hat{\phi}(0) = \check{\phi}(\xi_1).$$

Since there is no decay at all in the ξ_2 direction, this is not in L^p for any $p < \infty$.

Now how does the unit circle differ? We have

$$\lim_{\delta \to 0} \delta^{-1}\psi\left(\delta^{-1}(|\xi| - 1)\right),$$

and we have the estimate

$$\left|\left[\psi\left(\delta^{-1}(|\xi| - 1)\right)\right](x)\right| \leq C\delta|x|^{-\frac{n-1}{2}}\check{\psi}(\delta|x|).$$

107

Thus we expect that in the limit,

$$\left|\widehat{d\sigma}(x)\right| \leq C|x|^{-\frac{n-1}{2}},$$

which is indeed true. What we are lacking is a geometric understanding of the estimate on $\check{\psi}$. The computations in 5.2 started with the asymptotics for Bessel functions, and Bessel functions showed up again when we used spherical symmetry to compute $\widehat{d\sigma}$. The only problem with such formulae is that they average together too much; we want to localize. We start with a geometric understanding of the Fourier transform of an annulus in $\mathbb{R}^2$.

6.8 BUMPS REVISITED: Our problem is that the Fourier transform is ideally suited to viewing $\mathbb{R}^2$ as a product of one-dimensional spaces; it is easy to compute the transforms of rectangles; harder to do spheres. Thus our computation will be much easier if we decompose an annulus into rectangles. We use a partition of unity to decompose it into annular segments R_j, each of which has size $\sqrt{\delta} \times \delta$; this gives us $\frac{2\pi}{\sqrt{\delta}}$ pieces. If v_j denotes the unit vector whose tip is in the center of R_j, then $v_j = (\cos\theta_j, \sin\theta_j)$ where $\theta_j \approx \frac{j\sqrt{\delta}}{2\pi}$. We may view the rectangles as being translates of rectangles R_j^0 centered at the origin:

$$\chi_{R_j}(\xi) = \chi_{R_j^0}(\xi - v_j);$$

$$\check{\chi}_{R_j}(x) = e^{2\pi i v_j x}\check{\chi}_{R_j^0}(x).$$

To understand $\check{\chi}_{R_j^0}(x)$, we think of the integration

$$\int_{R_j^0} e^{2\pi i x \xi} d\xi$$

as a process of averaging certain waves against the rectangle. In two dimensions, these are plane waves, travelling in the direction $\frac{x}{|x|}$, with frequency $|x|$. Which directions and which frequencies contribute to the Fourier transform of $\chi_{R_j^0}$?

If the direction is parallel to v_j, that is, if $x = v_j|x|$, then the waves will average against an interval of thickness δ, so that the proper frquency for such an average to be non-zero is $|x| = \delta^{-1}$. Then the average gives the size of the rectangle, which is $\delta^{\frac{3}{2}}$, and therefore this frequency and direction contribute $e^{2\pi i x v_j}\delta^{\frac{3}{2}}$. But $x = v_j|x|$, $|x| = \delta^{-1}$, so that we get a contribution of

$$\frac{e^{2\pi i|x|}}{|x|^{\frac{3}{2}}}.$$

Now, which other directions contribute to the transform? At π, we get the same size contribution as at 0, but with a different phase, so the sum gives us a cosine. At a direction perpendicular to v_j, a wave coming in sees a thickness $\sqrt{\delta}$, so it selects a frequency $|x| = \delta^{-\frac{1}{2}}$, where $x = |x|w_j$; $v_j \cdot w_j = 0$. The contribution to the Fourier transform of such a wave is $e^{2\pi i x v_j}\delta^{\frac{3}{2}} = \frac{1}{|x|^3}$. Since this is much smaller, we take

108

the intuition that any directions not parallel to v_j can be ignored, though of course this is very simplistic. All in all,

$$\check{\chi}_{R_j}(x) = \frac{e^{2\pi i|x|}}{|x|^{\frac{3}{2}}}.$$

Now to compute $\check{\psi}_\delta$, we need to sum the contributions from the R_j. They will interact with each other on the x side, and we need to control this. Intuitively, we think of R_j^0 as a rectangle in the v_j and w_j directions, so that the Fourier transform is again a rectangle, S_j, with size $\delta^{-\frac{1}{2}}$ in the w_j direction and size δ^{-1} in the v_j direction. We claim that these rectangles essentially do not overlap. The S_j are at a distance $|x| = \delta^{-1}$ from the origin, and the angle between the successive rectangles is $C\delta^{\frac{1}{2}}$. Thus, between centers, there is a distance of

$$\delta^{-1}\delta^{\frac{1}{2}} = \delta^{-\frac{1}{2}}$$

in the w_j direction. Thus S_j and S_{j+1} essentially do not intersect, and, as j varies, the entire annulus is covered with the S_j. We conclude that

$$\check{\psi}_\delta(x) = \frac{\cos 2\pi|x|}{|x|^{\frac{3}{2}}} \quad |x| \approx \delta^{-1}.$$

We can get $\check{\chi}_D$ by adding up $\psi_{2^{-j}}$, or $d\check{\sigma}$ by taking a limit.

6.9 THEOREM. *If* $\lambda > \frac{n-1}{2(n+1)}$, *then* T_λ *is bounded on* $L^p(\mathbb{R}^n)$ *for the optimal range*

$$\frac{2n}{n+1+2\lambda} < p < \frac{2n}{n-1-2\lambda}.$$

PROOF: We decompose $K_\lambda = \sum K_j$ as before, but instead of interpolating, we use the philosophy that $\hat{K}_j = \psi_\delta \star \mu_\lambda$ behaves as though it were concentrated on an annulus of thickness δ. Then 6.6 controls restriction to such annuli;

$$\|K_j \star f\|_2^2 = \|\hat{K}_j \hat{f}\|_2^2$$

$$= \int_{1-\delta < |\xi| < 1+\delta} |\mu_\lambda(\xi)|^2 \left|\hat{f}(\xi)\right|^2 d\xi$$

$$\leq \delta^{\frac{1}{2}+\lambda} \left(\frac{1}{\delta} \int_{1-\delta}^{1+\delta} \int_{S^{n-1}} \left|\hat{f}(\sigma)\right|^2 d\sigma dr \right)$$

$$\leq C\delta^{\frac{1}{2}+\lambda} \|\hat{f}|_{S^{n-1}}\|_2 \leq C\delta^{\frac{1}{2}+\lambda} \|f\|_p^2.$$

There is a slight problem here: we need to control not $\|K_j \star f\|_2$ but $\|K_j \star f\|_p$, on the other hand, we can't get to the Fourier transform side to use restriction unless we start on L^2. We solve this problem with a great new trick.

We decompose $\mathbb{R}^n$ into disjoint cubes Q of diagonal length 2002^j, and we let $\tilde{Q}$ be the cube concentric with Q but of twice the diagonal length. Now

$$K_j \star f = \sum_Q K_j \star (f\chi_Q)$$

$$= \sum_Q \int_Q K_j(x-y)f(y)dy.$$

The integrand is zero unless there is a $y \in Q$ with $x - y \in support\ K_j$, i.e., $|x - y| < 2002^j = diameter\ Q$. Thus, $|x - center\ Q| \leq 2\ diameter\ Q$, and

$$support\ K_j \star (f\chi_Q) \subset \tilde{Q},$$

whence

$$K_j \star (f\chi_Q) = \sum [K_j \star (f\chi_Q)] \chi_{\tilde{Q}}.$$

Now for every $x \in \mathbb{R}^n$, there are at most 2^n cubes $\tilde{Q}$ with $x \in \tilde{Q}$, so that

$$\left| \sum [K_j \star (f\chi_Q)] \chi_{\tilde{Q}}(x) \right|^p$$

$$= \left| \sum_{j=1}^{2^{n+1}} a_j \right|^p \leq C_{p,n} \sum a_j^p$$

$$= \sum \left| \sum [K_j \star (f\chi_Q)] \chi_{\tilde{Q}} \right|^p ,$$

whence

$$\int |K_j \star f|^p \leq C_{p,n} \sum \int_{\tilde{Q}} |K_j \star (f\chi_Q)|^p .$$

We shall show that

$$\|K_j \star (f\chi_Q) \chi_{\tilde{Q}}\|_p^p \leq C 2^{-j(\frac{1}{2}+\lambda)} 2^{jn(\frac{1}{p}-\frac{1}{2})p} \|f\chi_Q\|_p^p$$

if $1 \leq p \leq \frac{2(n+1)}{n+3}$. Then, for this range of p,

$$\int |K_j \star f|^p \leq C \sum \int_{\tilde{Q}} |K_j \star f\chi_Q|^p$$

$$\leq C 2^{-\epsilon j} \sum \int_Q |f|^p = C 2^{-\epsilon j} \|f\|_p^p,$$

whence $\sum_p \|K_j\| < \infty$ if

$$\sum 2^{-j[(\frac{1}{2}+\lambda)+(\frac{1}{p}-\frac{1}{2})np]} < \infty.$$

To get this estimate, notice that

$$\int_{\tilde{Q}} |K_j \star f|^p \leq |\tilde{Q}|^{\frac{1}{r}} \left(\int |K_j \star f|^s \right)^{\frac{1}{s}} .$$

If we choose $ps = 2$, $\frac{1}{r} = 1 - \frac{p}{2}$ and $|\tilde{Q}|^{\frac{1}{r}} = 2^{j(1-\frac{p}{2})n}$. The theorem will follow from the estimate

$$\|K_j \star f\|_2 \leq C 2^{-j(\frac{1}{2}+\lambda)}.$$

Of course our intuition at the beginning is supposed to help; unfortunately ψ_δ is not really compactly supported, and $\hat{K}_j$ does not really live on an annulus. Thus there are contributions from 0 and ∞. The sneaky way to avoid the problem at zero is to define it out of existence. Let χ be in S with $\chi(r) = 0$ for $|r| > \frac{3}{4}$, $\chi(r) = 1$ for $|r| < \frac{1}{2}$. Let

$$p_\lambda(\xi) = \mu_\lambda(\xi) \chi(|\xi|)$$

$$q_\lambda(\xi) = \mu_\lambda(\xi)(1 - \chi(|\xi|))$$

$$k_\lambda = \check{\mu}_\lambda = \check{p}_\lambda + \check{q}_\lambda \equiv g_\lambda + b_\lambda.$$

Since μ_λ is smooth on the support of χ, $p_\lambda \in S$; $g_\lambda \in S$. Thus if $K_j = g_j + b_j$, $\|g_j\|_1 \leq c_k 2^{-kj}$ for every k, j. Thus the g's contribute error terms which we can handle. It suffices to start all over again with b_j. Zipping right by the first part, it is enough to prove the estimate:

$$\|b_j \star f\|_2 \le C 2^{-j(\frac{1}{2}+\lambda)} \|f\|_p.$$

We claim that if $|\xi| < \frac{1}{4}, |\hat{b}_j(\xi)| \le C_k 2^{-kj}$ for all k. Notice that

$$\hat{b}_j(\xi) = \int \psi_\delta(\xi - \eta)\mu_\lambda(\eta)\left[1 - \chi(|\eta|)\right] d\eta,$$

but $\mu_\lambda \in L^\infty$ and $\psi_\delta \in \mathcal{S}$, so that

$$|\psi_\delta(x)| \le C\delta^{-n}|\delta^{-1}x|^{-(n+k)},$$

and

$$\left|\hat{b}_j(\xi)\right| \le C\delta^k \int \frac{|1 - \chi(|\xi|)|}{|\xi - \eta|^{n+k}} d\eta.$$

Since $1 - \chi = 0$ unless $|\eta| > \frac{1}{2}$, $|\eta - \xi| \ge |\eta| - |\xi| \ge \frac{1}{2} - \frac{1}{4}$, the integral converges and our claim is true.

For the piece near infinity, we use the estimate

$$\int \left|\hat{f}(r, \sigma)\right|^2 d\sigma \le C r^{-\frac{n}{p'}} \|f\|_p^2.$$

Then

$$\|b_j \star f\|_2^2 = \int_0^\infty \int_{S^{n-1}} \left|\hat{b}_j(r)\right|^2 \left|\hat{f}(r, \sigma)\right|^2 r^{n-1} dr d\sigma$$

$$= \int_0^\infty \left|\hat{b}_j(r)\right|^2 \int_{S^{n-1}} \left|\hat{f}(r, \sigma)\right|^2 r^{n-1} dr d\sigma$$

$$\le C\|f\|_p^2 \int_0^\infty \left|\hat{b}_j(r)\right|^2 r^{n-1} r^{-\frac{n}{p'}} dr$$

$$\le C\|f\|_p^2 \left[C 2^{-kj} \int_0^{\frac{1}{4}} r^{\frac{n}{p}-1} dr + \int_{\frac{1}{4}}^\infty |\hat{b}_j(r)|^2 r^{n-1} dr\right]$$

$$\le C\|f\|_p^2 \left[C 2^{-kj} + \|\hat{b}_j\|_2^2\right].$$

But

$$\|\hat{b}_j\|_2^2 = \|b_j\|_2^2 \le 2\|K_j\|_2^2 + \|g_j\|_2^2$$

$$\le 2 \int_{100 \cdot 2^j}^{200 \cdot 2^j} r^{-\frac{n+1+2\lambda}{2}} r^{n-1} dr + C 2^{-kj} \le C 2^{-j(1+2\lambda)}.$$

INTRODUCTION): The analysis of the relation between Bochner-Riesz means and restriction is taken from C. Fefferman [23]. Notice that one needs to be quite careful in using this relationship, since, in fact, Bochner-Riesz means are bounded in the range $p > \frac{4}{3}$, whereas restriction fails precisely in this range.

PROPOSITION 6.1): These estimates and ideas originated in spectral synthesis theory; the radial restriction seems to be due to Stein, unpublished.

Since estimates on Bessel functions give rise to restriction, one can extend the radial restriction theorems in various ways, for example, by showing that if p is close to 1, $f \in L^p$ must have radial derivatives, or by showing that radial multipliers of L^p must have Holder continuity properties.

PROPOSITION 6.3): This was first proved by C. Fefferman and E. M. Stein, using arguments from spherical harmonics. The proof presented here is due to A. W. Knapp, unpublished.

PROPOSITION 6.5): The result for $p < \frac{2(n+1)}{n+3}$ comes from Tomas [51], with the proof given in the remark following the theorem. The strong result and the proof presented is due to Stein, unpublished.

ESSAY 6.7): The result that decrease of $\widehat{d\sigma}$ is equivalent to restriction is due to Stein; see C. Fefferman [22].

THEOREM 6.9): This result is due to C. Fefferman [22]; the proof presented here is due to Stein; see [23].

FURTHER REMARKS): The restriction theorem in $\mathbb{R}^n$ should be viewed as restricting functions in $\mathbb{R}^n$ to sets of positive co-dimension. For example, one might restrict to curves in $\mathbb{R}^3$ instead of to surfaces; this was first analyzed for general curves by Sjolin. There is an extensive literature on the subject, which we cannot hope to treat fairly. The best survey and best results as of this writing are due to Christ [6].

CHAPTER 7 THE MULTIPLIER PROBLEM FOR THE DISC

The purpose of this Chapter is to show that the characteristic function of the unit ball is not a multiplier of L^p except for $p = 2$, except in the known case of one dimension.

This ruins a lot.

SECTION 7.1 MEYER'S LEMMA

How can we guess a counterexample? Let's start on the Fourier transform side, and imagine a counterexample function f for which $\hat{f}$ is supported in a small rectangle. If the disc cuts the rectangle, the disc then acts on a small scale like the characteristic function of a half-plane in the direction normal to the disc. Since characteristic functions of half-planes give bounded multipliers, this gets us nowhere.

Fortunately, we already understand how discs differ from straight lines; it is all in how different rectangles pile up in different directions. What we should be doing is looking at lots of rectangles, $\hat{f}_j$, where the $\hat{f}_j$ are supported on disjoint rectangles R_j, which cover the edge of the disc. On each R_j the disc acts like a half plane, and all we have to do is analyze the overlap amongst the various f_j. The problem comes in doing a careful analysis of $\|\sum \check{\chi}_{R_j}\|_p$; it would be nice if we could take them to be disjoint. Instead, we use techniques from Chapter 4.

7.1 THEOREM. *If T_0 is bounded on some L^p, then for every sequence of functions $\{f_j\}$,*

$$\left\| \left(\sum |T_0 f_j|^2 \right)^{\frac{1}{2}} \right\|_p \le C \left\| \left(\sum |f_j|^2 \right)^{\frac{1}{2}} \right\|_p.$$

PROOF: Let $\vec{f} = (f_1, \ldots, f_k)$. Note that T_0 has a real valued convolution kernel, so it preserves real and imaginary parts of functions. We may assume then that the f_j are real valued. Let $\vec{v} = (v_1, \ldots, v_k)$ be a unit vector, and note that

$$\vec{v} \cdot \overline{T_0 f} = \sum v_j T_0 f_j = T_0(\vec{v} \cdot \vec{f}).$$

Thus

$$\int_{S^{n-1}} \int_{\mathbb{R}^n} \left| \vec{v} \cdot \overline{T_0 f}(x) \right|^p dx dv$$

$$\le \ _p\|T_0\|^p \int \int \left| \vec{v} \cdot \vec{f} \right|^p dx dv.$$

On the other hand,

$$\vec{v} \cdot \vec{f} = \|\vec{v}\| \|\vec{f}\| \cos \theta = \|\vec{f}\| \cos \theta,$$

so, after an interchange of integrals,

$$\int_{\mathbb{R}^n} \int_{S^{n-1}} |\cos \theta|^p dv \, \|\overline{T_0 f}(x)\|^p dx d\theta$$

$$\leq \ _p\|T_0\| \int \int |\cos\theta|^p dv \ \|\bar{f}\|^p d\theta dx.$$

Now for fixed $(f_1, \ldots, f_k)$, the vector $\bar{f}$ defines a north pole for S^{k-1}, and therefore the integral of $|\cos\theta|$ is independent of this vector. In particular, it is independent of x and pulls out as a constant on both sides of the inequality. Thus

$$\int \|\overline{T_0 f}\|^p dx \leq \ _p\|T_0\|^p \int \|\bar{f}\|^p dx,$$

or,

$$\int \left[\left(\sum |T_0 f_j|^2 \right)^{\frac{1}{2}} \right]^p dx$$

$$\leq \ _p\|T_0\|^p \int \left[\left(\sum |f_j|^2 \right)^{\frac{1}{2}} \right]^p.$$

To complete our intuitive arguments, we rigorize the idea that the disc acts like the characteristic function of a half-plane.

7.2 THEOREM. *Let v_j be unit vectors in $\mathbb{R}^n$; let $P_j = \{\xi | \xi \cdot v_j \geq 0\}$ and let H_j be the operator with multiplier χ_{P_j}. Then*

$$\left\| \left(\sum |H_j f_j|^2 \right)^{\frac{1}{2}} \right\|_p \leq \ _p\|T_0\| \left\| \left(\sum |f_j|^2 \right)^{\frac{1}{2}} \right\|_p.$$

PROOF: Let T_j^R be the operator corresponding to the multiplier

$$\chi_D \left(\frac{\xi - Rv_j}{R} \right);$$

let T_0^R correspond to $\chi_D \left(\frac{\xi}{R} \right)$. Then $T_j^R = J T_0^R J^{-1}$, where J is the isometry which multiplies f by $e^{2\pi i x v_j}$. Then

$$\left\| \left(\sum |T_j^R f_j|^2 \right)^{\frac{1}{2}} \right\|_p = \left\| \left(\sum |J T_0^R J^{-1} f_j|^2 \right)^{\frac{1}{2}} \right\|_p$$

$$= \left\| \left(\sum |T_j^R (J^{-1} f_j)|^2 \right)^{\frac{1}{2}} \right\|_p \leq \ _p\|T_0^R\| \left\| \left(\sum |J^{-1} f_j|^2 \right)^{\frac{1}{2}} \right\|_p$$

$$\leq \ _p\|T_0\| \left\| \left(\sum |f_j|^2 \right)^{\frac{1}{2}} \right\|_p.$$

We now take limits. The claim is that

$$\lim_{R \to \infty} \chi_D \left(\frac{\xi - Rv_j}{R} \right) = \chi_{P_j}(\xi).$$

Geometrically this is very clear; we are just saying that in the limit, a translated, dilated sphere becomes a plane (curvature 0). Analytically, the limit is 1 if $\lim \left| \frac{\xi - Rv_j}{R} \right| < 1$ and is zero otherwise. Thus, the limit is 1 if

$$\lim \frac{\xi - Rv_j}{R} \cdot \frac{\xi - Rv_j}{R} < 1$$

115

$$\lim \left(\frac{|\xi|^2}{R^2} - \frac{2Rv_j \cdot \xi}{R} + \frac{R^2|v_j|^2}{R^2} \right) < 1,$$

or, $-2v_j \cdot \xi + 1 < 1$.

Now if the f_j are in $\mathcal{S}$, $\chi_D \left(\frac{\xi - Rv_j}{R} \right) \hat{f}_j$ converge dominatedly to χ_{P_j}, so that $\lim T_j^R f_j = H_j f_j$ converges in L^2; a subsequence converges almost everywhere. Then

$$\left\| \left(\sum |H_j f_j|^2 \right)^{\frac{1}{2}} \right\|_p = \left\| \liminf_k \left(\sum |T_j^{R_k} f_j|^2 \right)^{\frac{1}{2}} \right\|_p$$

$$\leq \liminf_k \left\| \left(\sum |T_j^{R_k} f_j|^2 \right)^{\frac{1}{2}} \right\|_p$$

$$\leq {}_p\|T_0\| \left\| \left(\sum |f_j|^2 \right)^{\frac{1}{2}} \right\|_p.$$

7.3 Essay on Guessing a Counter-Example, Part II: In choosing f_j to constitute a counter-example, we expect that they will have Fourier transforms which are characteristic functions of rectangles. The rectangles best adapted to the disc are of size $\delta^{\frac{1}{2}} \times \delta$, with longest direction tangent to the disc at v_j. Then the f_j intuitively are characteristic functions of rectangles of size $N^{\frac{1}{2}} \times N$, with longest direction parallel to v_j. Since the transforms H_j of the previous result are in the direction v_j, we can expect that $H_j f_j \approx \frac{1}{2}$ on $\tilde{R}_j$, so that

$$\left\| \left(\sum \chi_{\tilde{R}_j} \right)^{\frac{1}{2}} \right\|_p \leq C \left\| \left(\sum \chi_{R_j} \right)^{\frac{1}{2}} \right\|_p.$$

Now the R_j are disjoint, so it is easy to compute $\left\| \left(\sum \chi_{R_j} \right)^{\frac{1}{2}} \right\|_p$, but the $\tilde{R}_j$ overlap, and the norm of $\sum \chi_{\tilde{R}_j}$ is quite complex. There is exactly one easy case: $p = 4$. Never mind that we already know the answer here; maybe we can learn something.

$$\left\| \left(\sum \chi_{\tilde{R}_j} \right)^{\frac{1}{2}} \right\|_4^4$$

$$= \int \left(\sum \chi_{\tilde{R}_j} \right)^2 = \sum_{j,k} \left| \tilde{R}_j \cap \tilde{R}_k \right|.$$

Now look at Figure 8. Let θ_j be the angle between v_j and the x-axis. Recall from the discussion in 6.7 that $\theta_j \approx j\delta^{\frac{1}{2}} = jN^{-\frac{1}{2}}$. Then the area of $\tilde{R}_j \cap \tilde{R}_k$ is the area of a parallelogram, which is the product of base and height. The dotted lines in the Figure are of length $N^{\frac{1}{2}}$, and $h = N^{\frac{1}{2}}$,

$$b = \frac{N^{\frac{1}{2}}}{\sin \theta}$$

$$= \frac{N^{\frac{1}{2}}}{\sin(\theta_j - \theta_k)} \approx \frac{N^{\frac{3}{2}}}{|j - k|}.$$

116

If $j = k$, the estimate $\left|\tilde{R}_j \cap \tilde{R}_k\right| = N^{\frac{3}{2}}$ is better; all in all,

$$\left|\tilde{R}_j \cap \tilde{R}_k\right| \geq \frac{N^{\frac{3}{2}}}{|j-k|+1}.$$

Then

$$\left\| \left(\sum \chi_{\tilde{R}_j}\right)^{\frac{1}{2}} \right\|_4^4 \geq \left(\sum_{j,k} \frac{N^{\frac{3}{2}}}{|j-k|+1}\right)$$

$$= C\left(N^{\frac{3}{2}}N^{\frac{1}{2}}\log N\right) = C\left(\log N\right)\left(\sum |R_j|\right).$$

In all,

$$_4\|T_0\| \geq C\left(\log N\right)^{\frac{1}{4}}.$$

As $\delta \to 0$, $N \to \infty$, and T_0 cannot be L^4 bounded.

In a way this trick does nothing more than compute the Fourier transform of an annulus, and show that such cannot be uniformly in L^p. If $p < 4$, we will need an entirely new idea, because the annuli behave very well in this range. What we do is let B be the set where all the $\tilde{R}_j$ overlap each other; we really only need to estimate $\left(\sum \chi_{\tilde{R}_j}\right)^{\frac{1}{2}}$ on B, because off B, the $\tilde{R}_j$ are disjoint, and therefore

$$\left(\sum \chi_{\tilde{R}_j}\right)^{\frac{1}{2}} \approx \sum \left(\chi_{\tilde{R}_j}\right)^{\frac{1}{2}},$$

whence, off B,

$$\left\| \left(\sum \chi_{\tilde{R}_j}\right)^{\frac{1}{2}} \right\|_p \approx \left\| \left(\sum \chi_{R_j}\right)^{\frac{1}{2}} \right\|_p$$

$$\approx \left(\sum |R_j|\right)^{\frac{1}{p}}.$$

On B, we need to come to terms with $\left(\sum \chi_{\tilde{R}_j}\right)^{\frac{p}{2}}$. Let $|B \cap \tilde{R}_j| = c|R_j|$, where, presumably, c is independent of j. Then

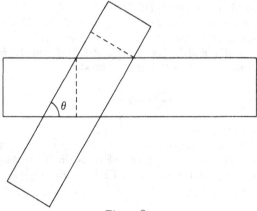

Figure 8

$$c\sum|R_j| = c\int_B \sum \chi_{\tilde{R}_j} \leq c\int_B \sum |H_j f_j|\chi_{\tilde{R}_j}$$

$$\leq |B|^{\frac{1}{r}}\| \left(\sum |H_j f_j|^2\right)^{\frac{1}{2}}\|_p^2 \leq |B|^{\frac{1}{r}} {}_p\|T_0\| \left(\sum |R_j|\right)^{\frac{2}{p}}.$$

Since $r = (\frac{p}{2})'$, $\frac{1}{r} = 1 - \frac{2}{p}$, and we get a lower bound

$$_p\|T_0\| \geq C \left(\frac{\sum |R_j|}{|B|}\right)^{1-\frac{2}{p}}.$$

If $p < 2$ this will not work, but if $p > 2$ we can get mileage out of this if $\frac{\sum |R_j|}{|B|}$ is large. This means that B has a substantially smaller measure than $\sum |R_j|$; on the other hand, $|\tilde{R}_j \cap B|$ has to be large - comparable to $|R_j|$. Thus the R_j must be carefully chosen so that the $\tilde{R}_j$ intersect in as small a set as possible.

SECTION 7.2 THE KAKEYA SET

7.4 THEOREM. *For each $k > k_0$ there is a set $E \subset \mathbb{R}^2$ and a collection of disjoint rectangles R_j such that, if $\tilde{R}_j$ is the pair of rectangles congruent to R_j but translated along R_j in the direction of the longest side of R_j, then*

$$\left|\tilde{R}_j \cap E\right| \geq \frac{1}{4}|R_j|$$

$$\frac{\sum |R_j|}{|E|} \geq C \log k.$$

PROOF: We get the set E by iterating a process known as "sprouting". The desire here is to construct triangles going in all sorts of directions, with a small common intersection. First we see how to do this with two triangles. Let ABC be the triangle in Figure 9. We construct new triangles AA'M and BB'M as in the Figure, by extending AC to AA', and by extending BC to BB'. Let M denote the midpoint of AB, and form the triangles AA'M, BB'M.

This process has constructed two new triangles going in different directions; we first show that no significant new area has been added. If b is the base length of AB, h_0 the height of ABC, and h_1 the heights of AA'M and BB'M, we claim that the union of the two new triangles has area at most

$$\frac{bh_0}{2} + 2\frac{bh_1^2}{h_0}.$$

Let E be the intersection of the lines AA', MB'; F the intersection of MA' and BB'. We bound the areas of the triangles EB'C and FA'C by the areas of triangles FEB' and FEA'. We concentrate on FEA'. Since FEA' is similar to MAA', and FEC is similar to BAC, we compute the height of FEA' as $h_1 + \frac{h_0 h_1}{h_0 + 2h_1} < 2h_1$, and we compute the base

$$|EF| = \frac{bh_1}{h_0 + 2h_1} < e\frac{bh_1}{h_0}.$$

118

The total area is bounded by $\frac{bh_1^2}{h_0}$. The same estimate applies to FEB'.

To construct the set E, we begin with an isosceles triangle of base b, height h_0, and sprout it with a height increment $\frac{h_0}{2}$. This results in two triangles, each of which is sprouted anew with height increment $\frac{h_0}{3}$. We continue to the kth step. We have 2^k triangles, each has base $b2^{-k}$, each has height $h_0 \sum \frac{1}{i} \approx h_0 \log k$. Let E be the union of all these triangles. What is the area of E? At the ith stage, we have 2^i triangles; fix one of them. It has height at least $h_0 \log i$, base at least $b2^{-i}$, and the height increment is $\frac{h_0}{i}$. The sprouting process adds an area increment of

$$2\frac{b}{2}^{-i}(\frac{h_0}{i})^2 h_0 \log i = 2\frac{bh_0 2^{-i}}{i^2 \log i}.$$

Since there are 2^i triangles, the total increment is $2\frac{bh_0}{i^2 \log i}$ and the area of E is bounded by

$$\frac{bh_0}{2} + 2 \sum bh_0 i^2 \log i < 3bh_0.$$

We now construct the rectangles R_j, corresponding to each of the 2^k triangles obtained above. Figure 10 sketches the idea. Let ABC be one of the sprouted triangles. Extend lines AC and BC to lines GAC and EBC. From point A draw a perpendicular to line DAC, until it meets EBC at F. The line AF is one side of R_j. The remaining side is AG, where G is chosen so that the length of AC is equal to the height of the sprout ABC.

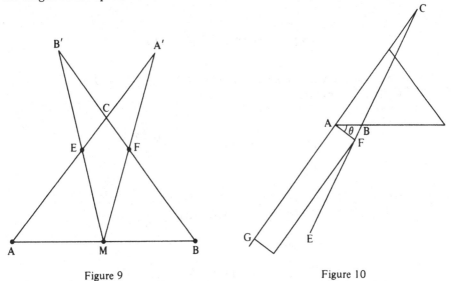

Figure 9   Figure 10

Clearly the R_j so constructed are disjoint, have length $h_0 \log k$ and width at most $b2^{-k}$. We also need a lower estimate on the width. Figure 10 indicates that the width is controlled by the eccentricity of the sprouted triangle. If θ is the angle between AF and AB, $\frac{AF}{AB} = \cos\theta$. Now $\theta = 90 - angle(AB, AC)$, and $angle(AB, AC) \geq 60$. since the sprouting process increases the angles of the corners,

and we started with an isosceles triangle. Thus $\theta < 30$ and $\cos\theta < \frac{\sqrt{3}AC}{2} = \frac{\sqrt{3}b2^{-k}}{2}$.
Then

$$\sum |R_j| = 2^k |R_0| \geq 2^k \left(\frac{\sqrt{3}b2^{-k}}{2} \right) (h_0 \log k)$$

and

$$\frac{\sum |R_j|}{|E|} \geq \frac{\sqrt{3}}{2} \frac{h_0 b \log k}{3h_0 b} \geq C \log k$$

as claimed.

Finally, we estimate

$$|\tilde{R}_j \cap E| \geq \frac{1}{4} |\tilde{R}_j|.$$

Refer to Figure 11. Clearly

$$|\tilde{R}_j \cap E| \geq |\tilde{R}_j \cap (ABC)| = |ABED|$$

$$= |AFED| - |ABF| \geq |AFED| - |AFG|;$$

in this estimate we used $AD = height\ ABC \leq AC$. Now, since D is not equal to E, $|AFED| \geq \frac{1}{2}|\tilde{R}_j|$. Finally,

$$|AFG| = \frac{1}{2}(AF)(FG) = (AF)^2 \frac{\tan\theta}{2}$$

$$\leq (AF)^2 \frac{\tan\frac{\pi}{6}}{2} \leq (AB)(AF) \frac{1}{2\sqrt{3}}$$

$$= \frac{b2^{-k}}{2\sqrt{3}}(AF) \leq h_0 \frac{\log k}{4}(AF) = \frac{1}{4}|\tilde{R}_j|.$$

Thus,

$$|\tilde{R}_j \cap E| \geq \frac{1}{2}|\tilde{R}_j| - \frac{1}{4}|\tilde{R}_j|.$$

120

7.5 THEOREM. T_0 is not bounded on $L^p(\mathbb{R}^n)$ if $n > 1$ and $p \neq 2$.

PROOF: Assume T_0 is bounded; we may asume that $p > 2$; we shall show that for each $k > k_0$,

$$_p\|T_0\| > C \, (\log k)^{\frac{1}{2} - \frac{1}{p}}.$$

Fixing k, we construct the Kakeya set E and rectangles R_j as in Theorem 7.4. If v_j is the direction of the longest side of R_j, we let $f_j = \chi_{R_j}$ and use Theorem 7.2 to obtain

$$\left\| \left(\sum |H_j f_j|^2 \right)^{\frac{1}{2}} \right\|_p \leq \; _p\|T_0\| \; \left\| \left(\sum |f_j|^2 \right)^{\frac{1}{2}} \right\|_p.$$

We shall show that for $x \in \tilde{R}_j$, $|H_j f_j(x)| \geq C$. It would then follow that

$$C \sum |R_j| = C \sum |\tilde{R}_j| \leq C \sum |\tilde{R}_j \cap E|$$

$$= C \sum \int_{\tilde{R}_j \cap E} \leq \sum \int_{\tilde{R}_j \cap E} |H_j f_j|^2$$

$$\leq \sum \int_E |H_j f_j|^2 = \int_E \sum |H_j f_j|^2 \leq |E|^{1 - \frac{2}{p}} \left(\int_E \left(\sum |H_j f_j|^2 \right)^{\frac{p}{2}} \right)^{\frac{2}{p}}$$

$$\leq |E|^{1 - \frac{2}{p}} \; _p\|T_0\| \; \left\| \left(\sum |H_j f_j|^2 \right)^{\frac{1}{2}} \right\|_p^2$$

$$\leq |E|^{1 - \frac{2}{p}} \; _p\|T_0\| \; \left\| \left(\sum |f_j|^2 \right)^{\frac{1}{2}} \right\|_p^2$$

$$\leq |E|^{1 - \frac{2}{p}} \; _p\|T_0\| \; \left(\sum |R_j| \right)^{\frac{2}{p}}.$$

Isolating $_p\|T_0\|$, we get

$$_p\|T_0\| \geq C \left(\frac{\sum |R_j|}{|E|} \right)^{1 - \frac{2}{p}} \geq C \, (\log k)^{1 - \frac{2}{p}}.$$

To complete the proof, we need to estimate $H_j f_j$. We remark that the estimates are rotation and translation invariant, whence we may assume $v_j = (0,1)$ and $R_j = (0,a) \times (0,b)$. Then

$$\chi_{P_j}(\xi_1, x_2) = \frac{1 - i(i \mathrm{sign} \xi_2)}{2}$$

so that

$$H_j = \frac{I \times I - iI \times H}{2}.$$

Since $f_j = \chi_{R_j}$ is real,

$$|H_j f_j| \geq \frac{1}{2} |(I \times H) f_j|.$$

But

$$(I \times H) f_j(x_1, x_2)$$

121

$$= \int \int \delta(x_1 - y_1)\chi_{(0,a)}(y_1)\frac{\chi_{(0,b)}(y_2)}{x_2 - y_2}dy_1\,dy_2$$

$$= \chi_{(0,a)}(x_1)\int \frac{\chi_{(0,b)}(y_2)}{x_2 - y_2}dy_2.$$

Now $\tilde{R}_j = (0,a) \times (-b,0) \cup (0,a) \times (b,2b)$, so that $(x_1, x_2) \in \tilde{R}_j$ implies $\chi_{(0,a)} = 1$ and $|x_2 - y_2| < 2b$, whence

$$|H_j f_j(x_1, x_2)| \geq \left|\int \frac{\chi_{(0,b)}(y_2)}{2b}dy_2\right| = \frac{1}{2}.$$

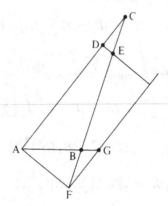

Figure 11

7.6 COROLLARY. T_0 is not bounded on $L^p(\mathbb{R}^n)$ if $n > 1$ and $p \neq 2$.

PROOF: If T_0 were bounded on L^p, we could restrict $\mu_0 = \chi_B$, $B = \{\xi|\ |\xi| < 1\}$ to planes in $\mathbb{R}^n$. By theorem 1.23 g, almost all of these would be multipliers of $L^p(\mathbb{R}^2)$. Now we choose planes perpendicular to the ξ_1 axis, and the restriction of μ_0 to these is a sequence of discs. At least one must be a multiplier of $L^l(\mathbb{R}^2)$, which is a contradiction.

7.7 ALTERNATE PROOF: In this essay we sketch a proof of 7.6 which is more in keeping with the intuition of how a counter-example should look.

We expected $\hat{f}_j$ to be the characteristic function of a rectangle of size $\delta \times \delta^2$, centered at the unit vector v_j; the counter-example functions f_j should be rectangles of size $N \times N^2$, multiplied by exponentials $e^{2\pi i v_j x}$.

7.8 LEMMA. For each $\gamma > 0$ and $k > k_0$ there is an N_k and a set E and 2^k disjoint rectangles $\{R_j\}$ satisfying

1) The longer side of each R_j has length N_k^2; the shorter side is bounded above and below by a constant times N_k.

2) $|E| \leq C(\log\log N_k)^{-1} \sum |R_j|$.

3) If $\tilde{R}_j$ is a translate of R_j by a distance $(1 + \gamma)N_k^2$ along the longest direction of R_j, then $|\tilde{R}_j \cap E| \geq C|R_j|$.

4) $\lim_{k\to\infty} N_k = \infty$.

PROOF: The proof is almost identical to that of 7.4. The starting triangle is given initial height $2^{2k}\log k$, and is sprouted with i^{th} height increment $\frac{2^{2k}\log k}{i}$. After k steps the triangles have height $N_k^2 = 2^{2k}\log^2 k$. The old R_j are constructed as before, and the new R_j are obtained by translating the old R_j by γN_k units away from E, and these are what we call R_j. Then $\bar{R}_j$ is our old $\tilde{R}_j$, so it enjoys the same estimates.

7.7 CONTINUED: To keep the computations manageable, we will use a simpler model; instead of using the correct

$$K_0(x) = \frac{\cos(2\pi|x| - \frac{3\pi}{4})}{|x|^{\frac{3}{2}}} + O\left(|x|^{-\frac{5}{2}}\right),$$

we use

$$K(x) = \frac{e^{i|x|}}{|x|^{\frac{3}{2}}}.$$

It is not possible to go from negative information about K to information about K_0, but a full treatment of K_0 uses the addition formulae for cos so many times that the unimportant error terms accumulate like flies. We'll do K.

As in the proof of 7.4, it is enough to show that $|K \star f_j(x)| \geq c$, $x \in \bar{R}_j$. We can do the computation intuitively from the analysis in the introduction to 6.1. Since $f_j = e^{iv_j \cdot y}\chi_{R_j}(y)$ and $x \in \bar{R}_j$, x is far from the support of f_j and

$$K \star f_j(x) \approx \frac{e^{i|x|}}{|x|^{\frac{3}{2}}}\hat{f}_j\left(\frac{x}{|x|}\right).$$

Now $\hat{f}_j$ is the characteristic function of a rectangle centered at v_j, times the homogeneity factor $N^{\frac{3}{2}}$, at least for all x such that $\frac{x}{|x|}$ is in the support of $\hat{f}_j$. This happens for $\left|\frac{x}{|x|} - v_j\right| < \frac{1}{N}$, which is valid when $x \in \bar{R}_j$. Then

$$|K \star f_j| \approx \frac{N^3}{|x|^{\frac{3}{2}}},$$

but $|x| \approx$ longest side of $R_j = \gamma N^2$, so that

$$|K \star f_j| \approx \frac{C}{\gamma^{\frac{3}{2}}}.$$

It is fairly easy to make this computation precise; we start with a rigorous version of the estimate in 6.1. Let w_j be a unit vector perpendicular to v_j; if $x \in \bar{R}_j$, $y \in R_j$, $\gamma > 2$, then

$$\gamma N^2 < |x - y| < 3\gamma N^2$$

$$\gamma N^2 \leq |(x - y) \cdot v_j| < 2\gamma N^2.$$

$$|(x - y) \cdot w_j| < \sqrt{2}N.$$

123

Then

$$|x - y| = \left([(x-y)\cdot v_j]^2 + [(x-y)\cdot w_j]^2 \right)^{\frac{1}{2}}$$

$$= |(x-y)\cdot v_j| \left(1 + \frac{[(x-y)\cdot w_j]^2}{[(x-y)\cdot v_j]^2} \right)^{\frac{1}{2}}.$$

Now using a Taylor series expansion of $\sqrt{1+x}$ about 0, and another for e^{ix} about 0, we get

$$e^{i|x-y|} = e^{i|(x-y)\cdot v_j|} + Error,$$

where

$$|Error| \leq Cmax \frac{[(x-y)\cdot w_j]^2}{[(x-y)\cdot v_j]}$$

$$\leq \frac{C_2 N^2}{\gamma N^2} = \frac{C}{\gamma}.$$

Now we choose the orientation of E so that $(x-y)\cdot v_j \geq 0$; then

$$|K \star f_j(x)|$$

$$\geq \left| \int \frac{e^{i(x-y)\cdot v_j}}{|x-y|^{\frac{3}{2}}} e^{iv_j \cdot y} \chi_{R_j}(y) dy \right| - \int_{R_j} \frac{Error}{|x-y|^{\frac{3}{2}}} dy.$$

The first term is

$$\int_{R_j} |x-y|^{\frac{3}{2}} dy \geq (3\gamma N^2)^{-\frac{3}{2}} |R_j| = C\gamma^{-\frac{3}{2}}.$$

The second term is bounded above by

$$\frac{C}{\gamma} \int_{R_j} |x-y|^{-\frac{3}{2}} dy$$

$$= \frac{C}{\gamma} \frac{C}{(2\gamma)^{\frac{3}{2}}} \frac{|R_j|}{N^3} \leq \frac{C}{\gamma^{\frac{5}{2}}}.$$

If γ is large enough, the second term is half the first, and

$$|K \star f_j| \geq \frac{C}{\gamma^{\frac{3}{2}}}.$$

7.1): Theorem 7.1 is taken from Zygmund [57]; the proof here does not work for operators T which mix real and imaginary parts of f. Paley showed how complex Gaussians may be used to establish the result for unpleasant T. Since we are really dealing with an Euclidean norm on $(f_1, f_2, \ldots, f_k)$, the whole subject is referred to as "vector-valued inequalities".

7.2): Theorem 7.2 is due to Yves Meyer, and the result is usually referred to as "Meyer's Lemma". It can be applied to sets besides the disc; in fact, if C is a convex set in $\mathbb{R}^2$ with normals in N distinct directions, a Meyer lemma for Hilbert transforms in these directions will apply. This will give us lower bounds for the L^p - multiplier norm of the characteristic function of a polygon with N sides. Similarly, a Meyer lemma can be used to show that L^p boundedness of the multiplier $\left|1 - |\xi|^2\right|^{it}$ leads to vector-valued inequalities for the convolution kernels $K_j(x) = |x \cdot v_j|^{-1-it}$.

The most far-reaching application of Meyer's Lemma is in the theory of A. Cordoba and R. Fefferman, relating maximal functions and multipliers.

7.4): The Kakeya set was originally constructed to solve a geometric problem in the plane. Consider a line segment I of length one; let E be a set. We are allowed to rotate and translate I as long as I remains inside E. What is the minimal area of E in which I can be rotated a full 360°? The Kakeya set can be used to show that $|E|$ can be made less than ϵ for every ϵ. See de Guzman [18] for an excellent exposition. Generalizations are discussed in Cunningham [17].

Variants of the Kakeya set can be used to analyze other geometric-analytic problems in $\mathbb{R}^2$, particularly the differentiation of integrals. The problem is this: construct a maximal function for functions on $\mathbb{R}^2$ based on averaging over squares or balls; one dimensional ideas prove weak L^1 and L^p bounds. But if you try to average over rectangles in arbitrary directions, Kakeya set techniques show that the *supremum* is infinity. See again de Guzman [18].

7.4): This result is due to C. Fefferman. Clearly the same ideas generalize to any multiplier which has a discontuity along a contiuum of directions; in particular, to jump discontinuities of radial functions. The case of oscillatory discontinuities is open.

7.7): The alternative proof here is derived from the sketch in C. Fefferman [20]; see [32]. The point of computational techniques is that they apply even when the Meyer lemma does not; see Ruiz [11]

CHAPTER 8 THE CORDOBA MULTIPLIER THEOREM

The purpose of this chapter is to prove the $L^p(\mathbb{R}^2)$ boundedness of Bochner-Riesz means, in the optimal range. Many individuals gave many different proofs; the first is due to L. Carleson and P. Sjolin [5]. In this chapter we shall focus on a fairly recent proof due to A. Cordoba [10]; it combines the main themes of maximal function and the standard model.

SECTION 8.1 ALMOST ORTHOGONALITY

The standard model of singularities suggests that we need a good estimate on $\phi_\delta(\xi) = \phi\left(\delta^{-1}(|\xi|-1)\right)$ and its multiplier properties. For the counterexample functions of Essay 7.3, $T_\delta f_j = T_0 f_j$, and we therefore expect that $_4\|T_\delta\| \geq C|\log\delta|^{\frac{1}{4}}$. If in fact this lower bound is close to the true operator norm, then the standard model suggests that $\mu_\lambda = \sum 2^{-\lambda j}\phi_{2^{-j}}$, so that

$$_4\|T_\lambda\| \leq \sum 2^{-\lambda j} \,_4\|T_{2^{-j}}\| \leq C\sum 2^{-\lambda j}j^{\frac{1}{4}} < \infty$$

when $\lambda > 0$. This would prove the L^4 boundedness of all T_λ and then duality and interpolation would prove the optimal result everywhere else, as sketched in Figure 12.

The computations in 7.3 suggest that the ϕ_δ ought to be decomposed into small rectangles of size $\delta \times \delta^{\frac{1}{2}}$; after controlling the norms of each of these, we would have to prove some sort of converse to the Meyer Lemma, isolating the effects of each rectangle. So we begin by building orthogonality results.

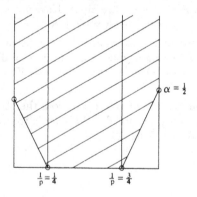

Figure 12

8.1 THEOREM. *Let I_j be intervals of equal length and disjoint interior whose union is all of $\mathbb{R}$. Let P_j be the multiplier operator with multiplier χ_{I_j}. If $2 \leq p < \infty$, then*

$$\| (|P_j f|^2)^{\frac{1}{2}} \|_p \leq C\|f\|_p.$$

PROOF: The grid of intervals may be translated until they are all of the form $I_j = (ja, (j+1)a)$; this alters f by an isometry. A dilation gives $(j, j+1)$ and another translation $(j - \frac{1}{2}, j + \frac{1}{2})$. We conclude that the norm in the theorem is independent of the size of the intervals.

In proving the Littlewood-Paley inequalities, we needed: a) a smooth version of the result, b) a multiplier theorem to prove a), and c) an orthogonalization proceedure. To construct the multiplier result, we let ψ be a smooth function satisfying $\psi = 1$ on I_0; $\psi = 0$ off $\tilde{I}_0$. Let T_j be the multiplier operator corresponding to $\psi(\xi - j)$; note that $P_j f = P_j T_j f$. Since the Hormander-Mihilin theorem does not apply to $\sum T_j$, we need a new trick, which will turn out to handle our orthogonality needs at the same time. Let

$$\mu(\xi) = \sum e^{2\pi i j \theta} \psi(\xi - j); \ k = \check{\mu}.$$

Then

$$\int_{I_0} |k \star f(x)|^2 \, d\theta = \int_{I_0} \left| \sum e^{2\pi i j \theta} T_j f(x) \right|^2 d\theta = \sum |T_j f(x)|^2$$

by the Plancherel theorem for I_0. The question, though, is about L^p behaviour of convolution with k; for this we use the Poisson summation formula (*cf.* Stein and Weiss [50] Chapter 7 Section 2).

If $f \in L^1(\mathbb{R})$, $\theta \in I_0$, then

$$\sum \hat{f}(j) e^{2\pi i j \theta} = \sum f(\theta + j);$$

here $\hat{f}(j) = \int e^{-2\pi i j y} f(y) dy$. In our case, we need $\hat{f}(j) = \psi(\xi - j)$, so we set

$$f(x) = \int e^{2\pi i y x} \psi(\xi - y) dy = e^{2\pi i x \xi} \check{\psi}(x).$$

Since $\psi \in S$, $f \in L^1$. Then

$$\sum \psi(\xi - j) e^{2\pi i j \theta} = \sum e^{2\pi i (j + \theta) \xi} \check{\psi}(\theta + j),$$

and

$$k(x) = \sum \left[e^{2\pi i (j + \theta) \xi} \right] (x) \check{\psi}(\theta + j)$$

$$= \sum \delta_{j+\theta}(x) \check{\psi}(j + \theta).$$

Now the L^p operator norm of convolution with k is given by the total variation norm of k;

$$\|k\| \leq \sum \left| \check{\psi}(j + \theta) \right| \|\delta_{j+\theta}\|$$

$$\leq \sum \left| \check{\psi}(j + \theta) \right| \leq C \sum \left((j + \theta)^2 + 1 \right)^{-1} \leq C.$$

This controls the multiplier norm; finally we need to get from L^2 to L^p; we will use Holder's inequality instead of independence. Then

$$\int \left(\sum |P_j f|^2 \right)^{\frac{p}{2}} = \int \left(\sum |P_j T_j f|^2 \right)^{\frac{p}{2}}$$

127

$$\leq C \int \left(\sum |T_j f|^2 \right)^{\frac{p}{2}},$$

from Proposition 4.12. This in turn is equal to

$$C \int \left(\int_{I_0} |k \star f|^2 d\theta \right)^{\frac{p}{2}} dx$$

$$= C \int (\|k \star f\|_{2,\theta})^p \, dx \leq \int (\|k \star f\|_{p,\theta})^p$$

$$= C \int \int_{I_0} |k \star f|^p d\theta dx = C \int_{I_0} \int_{\mathbb{R}} |k \star f|^p dx d\theta$$

$$\leq C \|k\| \int_{I_0} \int |f|^p dx d\theta = C \|f\|_p^p.$$

8.2 COROLLARY. Let $B_j = \{(\xi_1, \xi_2) | \xi_2 \in I_j\}$, and let S_j be the operator with multiplier χ_{B_j}. Then for $2 \leq p < \infty$,

$$\left\| \left(\sum |S_j f|^2 \right)^{\frac{1}{2}} \right\|_p \leq C \|f\|_p.$$

SKETCH OF PROOF: Since S_j acts like the identity in the ξ_1 direction, the result is a product of Theorem 8.1 and the identity. Formally, for each x_1 one would define a function $g(x_2) = f(x_1, x_2)$, and note that for almost all x_1, $g \in L^1$ and $(S_j f)(x_1, x_2) = (P_j g)(x_2)$. And so on.

REMARK: The regular distribution of the I_j was not essential in the proof of 8.1; all we really was that $\psi_j(\xi) = \psi(\xi - j)$ and that $\chi_{I_j} \psi_j = \chi_{I_j}$. This means that the I_j can be fudged and we will still have a theorem. Certainly we can drop every other j, and thus assume that $distance\,(I_j, I_{j+1}) \geq c > 0$. But we must also have that the lengths of the I_j are comparable to their separation and to each other;

$$c_1 \leq d(I_j, I_{j+1}) \leq c_2$$

$$c_1 \leq |I_j| \leq c_2.$$

We actually have an application in mind; let

$$C_j = \{(r, \theta) | \, j \leq \delta^{-\frac{1}{2}} \theta \leq j + 1\},$$

and let I_j be the projection of C_j onto the ξ_2 axis. Then

$$|I_j| \approx \delta^{\frac{1}{2}} \sin \left(\frac{2\pi j}{\delta^{\frac{1}{2}}} \right),$$

so that the lengths of the I_j are not comparable. If we divide the I_j into four groups, each of which lies in a cone about the $\frac{\pi}{4}$ lines, and we treat the intervals in each cone by a separate square function, then within the cones we can guarentee comparable size.

128

8.4 ALMOST ORTHOGONALITY AND CURVATURE: Unfortunately, 8.2 is not good enough; to get multiplier theorems, we need inequalities of the form

$$\|T_\delta f\|_4 \leq \left\| \left(\sum |S_j T_\delta f|^2\right)^{\frac{1}{2}} \right\|_4 \leq \left\| \left(\sum |P_j f|^2\right)^{\frac{1}{2}} \right\|_4;$$

Corollary 8.2 only allows us to go from $\sum |P_j f|^2$ to $|f|$, and so we need some more orthogonality. To see what we need, let C_j denote the cones

$$\{(r, \theta)|j \leq \delta^{-\frac{1}{2}}\theta \leq j + 1\},$$

and let $\mu_j = \chi_{C_j} \phi_\delta$. As above, we only treat C_j which lie in the cone $0 \leq \theta \leq \frac{\pi}{4}$. If T_j is the operator corresponding to μ_j, let $T^0 = \sum T_j$. Let I_j denote the projections of the support of the μ_j onto the ξ_2 axis, and let S_j be the operator with multiplier $\chi_{I_j}(\xi_2)$. Then $S_j T_j = T_j$, and the estimate

$$\left\| \sum T_j f \right\|_4 = \left\| \left(\sum |T_j f|^2\right)^{\frac{1}{2}} \right\|_4$$

would then imply that

$$\|T^0 f\|_4 = \left\| \sum T_j f \right\|_4 \leq C \left\| \left(\sum |T_j f|^2\right)^{\frac{1}{2}} \right\|_4$$

$$= \left\| \left(\sum |S_j T_j f|^2\right)^{\frac{1}{2}} \right\|_4.$$

We would like to control this as

$$C \left\| \left(\sum |S_j f|^2\right)^{\frac{1}{2}} \right\|_4,$$

which we can handle using 8.2. So we need two inequalities:

$$\left\| \sum T_j f \right\|_4 \leq C \left\| \left(\sum |T_j f|^2\right)^{\frac{1}{2}} \right\|_4$$

$$\left\| \left(\sum |S_j T_j f|^2\right)^{\frac{1}{2}} \right\|_4 \leq C \left| \left(\sum |S_j f|^2\right)^{\frac{1}{2}} \right\|_4.$$

To analyze the first, we take 4th powers: we need

$$\int \left| \sum T_j f \right|^4 \leq C \int \left(\sum |T_j f|^2\right)^2$$

$$\int \left| \left(\sum T_j f\right)\left(\sum T_k f\right) \right|^2 \leq C \int \sum |T_j f|^2 |T_k f|^2.$$

Leave aside the integrals for a second; we are asking that

$$\left(\sum |a_j|\right)^2 \leq C \sum |a_j|^2.$$

129

This is very peculiar; all the true inequalities, like $(a+b)^2 \geq a^2 + b^2$, go in the other direction. Of course it is true that $(a+b)^2 \leq 2(a^2 + b^2)$, but this gets less true the more terms there are inside the square. Thus the integration is critical: we had better hope that if x is in the support of one of the $T_j f T_k f$, it is not in the support of the others. In short, we hope that the $T_j f T_k f$ are orthogonal on L^2.

Of course, this is L^2, and orthogonality ought to mean something on the Fourier transform side.

$$\int \sum |T_j f T_k f|^2 = \int \left| \sum (T_j f T_k f) \right|^2$$

$$= \int \left| \widehat{(T_j f)} \star \widehat{(T_k f)} \right|^2$$

$$= \int \left| \sum \left(\mu_j \hat{f} \right) \star \left(\mu_k \hat{f} \right) \right|^2 .$$

If A_{jk} denotes the support of

$$\left(\mu_j \hat{f} \right) \star \left(\mu_k \hat{f} \right),$$

we shall show that for all ξ, $\sum \chi_{A_{jk}}(x) \leq 9$, that is, no ξ is in more than 9 of the A_{jk}. This is a substitute for complete disjointedness, implying orthogonality. It is just as good, because

$$\left| \sum \left(\mu_j \hat{f} \right) \star \left(\mu_k \hat{f} \right) \right|$$

$$= \left| \sum \left(\mu_j \hat{f} \right) \star \left(\mu_k \hat{f} \right) \chi_{A_{jk}} \right|$$

$$\leq \left(\sum |\chi_{A_{jk}}|^2 \right)^{\frac{1}{2}} \left(\sum |\mu_j \hat{f} \star \mu_k \hat{f}|^2 \right)^{\frac{1}{2}}$$

$$\leq 3 \left(\sum |\mu_j \hat{f} \star \mu_k \hat{f}|^2 \right)^{\frac{1}{2}} .$$

Squaring and integrating,

$$\|T_0 f\|_4^4 = \int \left| \sum \left(\mu_j \hat{f} \right) \star \left(\mu_k \hat{f} \right) \right|^2$$

$$\leq 9 \int \sum | \left(\mu_j \hat{f} \right) \star \left(\mu_k \hat{f} \right) |^2$$

$$= 9 \int \sum |T_j f|^2 |T_k f|^2 = 9 \int \left(\sum |T_j f|^2 \right)^2$$

$$= 9 \| \left(\sum |T_j f|^2 \right)^{\frac{1}{2}} \|_4^4 ,$$

and the first part of the analysis is complete. To show that $\sum \chi_{A_{jk}} \leq 9$, we remark that

$$A_{jk} = support \; (\mu_j \star \mu_k) = support \; \mu_j + support \; \mu_k.$$

The task ahead of us is geometric: we have to analyze the intersections of the A_{jk}, sums of sectors of an annulus. The actual geometric picture is complex; Figure 13 represents what the picture would look like if the μ_j were supported on a circle instead of an annulus; in Figure 14 we sketch an actual A_{jk}. To prove disjointednes, the A_{jk} must be separated in the angular and radial directions. If v_j is the unit vector at the center of the support of μ_j, then A_{jk} is a set symmetric about the direction $v_j + v_k$. There are at most $C\delta^{-\frac{1}{2}}$ such pairs of directions; along each, A_{jk} is contained in a ball of radius $2\delta^{\frac{1}{2}}$. We take the center of the ball to be $v_j + v_k$. Since these vectors are equally spaced, a cone about $v_j + v_k$ isolates the A_{jk} from each other; we may conclude that $A_{jk} \cap A_{j'k'}$ not equal to ϕ implies $|(j+k)-(j'+k')| < 2$. This accomplishes isolation in angular directions.

For the radial directions, notice that A_{jk} is centered at

$$\left(1 - \frac{\delta}{2}\right)(\cos\theta_j + \cos\theta_k, \sin\theta_j + \sin\theta_k);$$

to within an accuracy δ, $\theta_j = j\delta^{\frac{1}{2}}$. The distance between the centers of the A_{jk} is therefore the length of the vector

$$\left(1 - \frac{\delta}{2}\right)(\cos\theta_j - \cos\theta_{j-1} + \cos\theta_k - \cos\theta_{k+1}, \ldots),$$

which is

$$2\left(1 - \frac{\delta}{2}\right)(\sin\left(\frac{\theta_j + \theta_{j-1}}{2}\right)\sin\left(\frac{\theta_j - \theta_{j-1}}{2}\right)$$

$$+ \sin\left(\frac{\theta_k + \theta_{k+1}}{2}\right)\sin\left(\frac{\theta_k - \theta_{k+1}}{2}\right), \ldots).$$

Since $\theta_j - \theta_{j-1} = \frac{\delta^{\frac{1}{2}}}{4}$ to an accuracy δ, the centers are spaced a distance $\delta^{\frac{1}{2}}$ apart, to accuracy δ.

The sets A_{jk} in the direction $v_j + v_k$ therefore constitute $\delta^{-\frac{1}{2}}$ sets, each of which is contained in a ball of radius $2\delta^{\frac{1}{2}}$, and whose centers are equidistributed along a line in the $v_j + v_k$ direction. Thus, no ball can overlap more than two others along a line, and no more than three others on adjacent lines. Thus no point of the plane can belong to more than 9 of the A_{jk}.

We now need to understand the inequality

$$\left\| \left(\sum |S_j T_j f|^2\right)^{\frac{1}{2}} \right\|_4 \leq \left\| \left(\sum |S_j f|^2\right)^{\frac{1}{2}} \right\|_4.$$

If we recall the proof of the Marcinkiewicz theorem, 4.14, we see that we need to get inequalities like

$$\int |S_j T_j f|^2 \, g \leq \int |S_{jj} f|^2 g^{\star}$$

131

where g^* is some sort of maximal function. Thinking along the lines of Chapter 7, $S_j T_j f$ acts like a Hilbert transform in the v_j direction. If M_j is the one dimensional maximal function in this direction, we would expect from Chapters 3 and 4 that

$$\int |S_j T_j f|^2 \, g \leq \int |S_j f|^2 \, [M_j(g^*)]^{\frac{1}{2}}$$

and therefore that

$$\int \sum |S_j T_j f|^2 \, g$$

$$\leq \int \sum |S_j f|^2 \sup_j \, [M_j(g^*)]^{\frac{1}{2}}$$

$$\leq \| \left(\sum |S_j f|^2 \right)^{\frac{1}{2}} \|_4^2 \, \| \sup_j \, [M_j(g^*)]^{\frac{1}{2}} \|_2;$$

the only problem remaining is to control the maximal function. This approach does succeed, but $S_j T_j$ really corresponds to μ_j, which is less singular than a Hilbert transform; if we take this into account, the maximal function will be easier to control. Intuitively, μ_j is the characteristic function of a rectangle of size $\delta \times \delta^{\frac{1}{2}}$; T_j therefore should act like averaging over a rectangle with dual size $\delta^{-1} \times \delta^{-\frac{1}{2}}$. This is the type of maximal function we need to control.

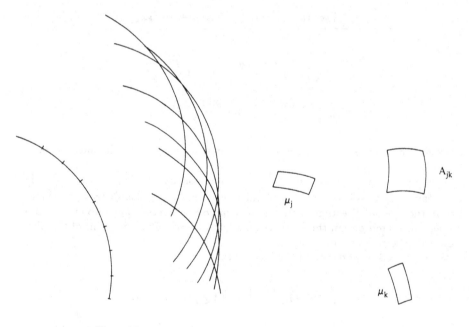

Figure 13   Figure 14

SECTION 8.2 THE CORDOBA MAXIMAL FUNCTION

8.5 DEFINITION. *Fix $N > 1$, $a > 0$. Let*

$$\mathcal{B} = \{R \mid R \text{ is a rectangle with dimension } a \times aN\}.$$

The Cordoba maximal function $\mathcal{C}$ is defined as

$$\mathcal{C}f(x) = \sup \frac{1}{|R|} \int_R |f(y)| dy.$$

The supremum is taken over all rectangles in $\mathcal{B}$ which contain x.

8.6 THEOREM.
$$\|\mathcal{C}f\|_2 \leq C \, (\log 6N)^{\frac{1}{2}} \, \|f\|_2.$$

PROOF: Quantities like $\|\sum \chi_{R_j}\|_2 = \sum |R_j \cap R_k|$ can be controlled as in Chapter 7, and this produces the log terms we need. Unfortunately, the process of forming $\mathcal{C}$ averages f over rectangles like the R_j; it does not generate χ_{R_j}.

A key new idea in the Cordoba result is that we can change from averaging over R_j to multiplying by χ_{R_j} by the process of forming adjoints: If $\mathcal{C}f = \int f \chi_R$;

$$\mathcal{C}^* f = \left(\int f \right) \chi_R.$$

Of course there are technical details here.

The first technical difficulty is that for each x, we have a rectangle R for which

$$\mathcal{C}f(x) \approx \frac{1}{|R|} \int_R f.$$

The dependence of R on x makes it impossible to compute adjoints. Therefore our first duty is to localize x, so that one rectangle will work for lots of x.

We begin by restricting the permitted angles of the rectangles in $\mathcal{B}$ to $(0, \frac{\pi}{4})$; the corresponding maximal function is still denoted $\mathcal{C}$. We really are looking for regions over which $\mathcal{C}$ is constant. But $\mathcal{C}$ is the supremum of convolution operators with kernels $|R|^{-1}\chi_R$; as in Chapter 7, $\mathcal{C}$ acts like a convolution kernel with compact support, and therefore acts independently on squares equal to the diameter of the support. Since all rectangles in $\mathcal{B}$ have size $a \times aN$, we divide $\mathbb{R}^2$ into a disjoint union of squares Q with side length aN; let $\tilde{Q}$ be the concentric squares with side length $6aN$. If $\mathcal{C}(f\chi_Q)(x) \neq 0$, there is an $R \in \mathcal{B}$ with $x \in R$ and

$$|R|^{-1} \int_R f(y)\chi_Q(y) dy > 0.$$

Therefore, $y \in R \cap Q$ and

$$|x - center \; Q| \leq |x - y| + |y - center \; Q| \leq diameter \; R + diameter \; Q$$

133

$$\leq 2\sqrt{2}aN \leq 3aN.$$

Thus, *support* $\mathcal{C}(f\chi_Q) \subset \tilde{Q}$. Since each $x \in \mathbb{R}^2$ is in a unique Q, and is in at most 30 of the $\tilde{Q}$,

$$\|\mathcal{C}f\|_2^2 = \|\mathcal{C}\left(\sum f\chi_Q\right)\|_2^2$$

$$\leq \|\sum \mathcal{C}(f\chi_Q)\|_2^2 = \|\sum \mathcal{C}(f\chi_Q)\chi_{\tilde{Q}}\|_2^2$$

$$\leq \sqrt{30}\sum \|\mathcal{C}(f\chi_Q)\|_2^2$$

$$\leq C\|(\log 3N)\sum \|f\chi_Q\|_2^2 = C(\log 3N)\|f\|_2^2;$$

the theorem is complete if we prove

$$\|\mathcal{C}(f\chi_Q)\chi_{\tilde{Q}}\|_2 \leq C(\log 3N)^{\frac{1}{2}}\|f\chi_Q\|_2.$$

To do this, we fix a cube Q and write f instead of $f\chi_Q$.

So far, we have made $\mathcal{C}f$ constant, that is, zero, off $\tilde{Q}$; now we chop $\tilde{Q}$ into pieces on which $\mathcal{C}$ is constant. Since each R is of size $a \times aN$, and the directions of the R are in a an angular region of size $\frac{\pi}{4}$, $\mathcal{C}$ will act like it has support size a in the direction perpendicular to the direction of R. Subdivide $\tilde{Q}$ into $36N^2$ squares Q_{ij} of size $a \times a$. We claim that for each Q_{ij} there is an $R_{ij} \in \mathcal{B}$ for which

$$\mathcal{C}f(x) \leq 2|R_{ij}|^{-1}\int_{R_{ij}}|f|$$

for all $x \in Q_{ij}$. If so, we could dominate

$$\mathcal{C}f(x) \leq 2\sum|R_{ij}|^{-1}\int_{R_{ij}}|f|$$

and then compute the adjoint of the right-hand side operator.

If $R \in \mathcal{B}$ and $R \cap Q_{ij} \neq \phi$, then Q_{ij} is contained in $\tilde{R}$ where $\tilde{R}$ is the rectangle concentric with R, having eight times the diameter. Then

$$|R|^{-1}\int_R|f| \leq 64|\tilde{R}|^{-1}\int_{\tilde{R}}|f|;$$

and for $x \in Q_{ij}$,

$$\mathcal{C}f(x) \leq \sup_{R\in\mathcal{B}}64|\tilde{R}|^{-1}\int_{\tilde{R}}|f| \leq C\sup_{R\in\mathcal{B}'}|R|^{-1}\int_R|f|,$$

where

$$\mathcal{B}' = \{R|Q_{ij} \subset R;\ direction R \in (0, \frac{\pi}{4}),\ size\ R = 8a \times 8aN\}.$$

134

But the supremum of $R \in \mathcal{B}'$ is constant on Q_{ij}, so we can find an $R_{ij} \in \mathcal{B}$ for which $\tilde{R}_{ij} \in \mathcal{B}'$ and for which

$$\frac{1}{2}\mathcal{C}f(x) \le 64|\tilde{R}_{ij}|^{-1} \int_{\tilde{R}_{ij}} |f|$$

for all $x \in Q_{ij}$.

What we've done so far is to localize the maximal function; we now show that the expected estimate on intersections of rectangles is enough to finish the proof. So, assume we've chosen f and obtained the $\tilde{R}_{ij}$. Assume that we can prove

$$\left| \tilde{R}_{ij} \cap \tilde{R}_{i'j'} \right| \le \frac{CaN}{|j - j'| + 1}.$$

We now define an operator $T : L^2(Q) \to L^2(\tilde{Q})$ by

$$Tg(x) = \sum |\tilde{R}_{ij}|^{-1} \int_{R_{ij}} g(y)dy \, \chi_{Q_{ij}}(x).$$

Notice that for our fixed f, $\mathcal{C}f(x) \le 2T\left(|f|\right)(x)$, so that the theorem is proved once we get L^2 bounds on T, which follow from bounds on T^*. But

$$\int T^*h(y)g(y)dy = \int h(x)Tg(x)dx$$

$$= \sum |\tilde{R}_{ij}|^{-1} \int_{R_{ij}} g(y)dy \int h(x)\chi_{Q_{ij}}(x)dx$$

$$= \int \left(\sum |\tilde{R}_{ij}|^{-1} \int_{Q_{ij}} h(x)dx \chi_{\tilde{R}_{ij}}(y) \right) g(y)dy.$$

Thus

$$T^*h(y) = \sum |\tilde{R}_{ij}|^{-1} \int_{Q_{ij}} h(x)dx \chi_{\tilde{R}_{ij}}(y)$$

$$= (64a)(aN)^{-1} \sum \int_{Q_{ij}} h(x)dx \chi_{\tilde{R}_{ij}}(y) = CN^{-1} \sum a_{ij} \chi_{\tilde{R}_{ij}}(y).$$

Computing L^2 norms,

$$\int |T^*h|^2 = N^{-2} \sum a_{ij} a_{i'j'} \left| \tilde{R}_{ij} \cap \tilde{R}_{i'j'} \right|$$

$$\le Ca^2 N^{-1} \sum a_{ij} a_{i'j'} \left(|j - j'| + 1 \right)^{-1}$$

$$= Ca^2 N^{-1} \sum_{i,i'} \sum_j a_{ij} \sum_{j'} a_{i'j'} \left(|j - j'| + 1 \right)^{-1}$$

135

$$\leq C a^2 N^{-1} \sum_{i,i'} \left[\sum_j a_{ij}^2 \right]^{\frac{1}{2}} \left[\sum_j \left(\sum_k a_{i',j-k} \left(|k'| + 1 \right)^{-1} \right)^2 \right]^{\frac{1}{2}}$$

$$\leq C a^2 N^{-1} \sum_{i,i'} \left[\sum_j a_{ij}^2 \right]^{\frac{1}{2}} \left[\sum \left(|k'| + 1 \right)^{-1} \right] \sum_{i'} \left[\sum_{j'} a_{i'j'}^2 \right]^{\frac{1}{2}}$$

$$= C a^2 N^{-1} \log 6N \left(\sum_i \left[\sum_j a_{ij}^2 \right]^{\frac{1}{2}} \right)^2$$

$$\leq C a^2 N^{-1} \log 6N \left(\left[\sum_j a_{ij}^2 \right]^{\frac{1}{2}} \right)^2 \left(\sum_i \left[\sum_j a_{ij}^2 \right]^{\frac{1}{2}} \right)^{\frac{1}{2}}$$

$$= C a^2 \log 6N \sum_i a_{ij}^2 = C a^2 \log 6N \sum \left(\frac{1}{|Q_{ij}|} \int_{Q_{ij}} h \right)^2$$

$$\leq C a^2 \log 6N \sum \frac{1}{|Q_{ij}|} \int_{Q_{ij}} h^2 = C \log 6N \sum \int_{Q_{ij}} h^2$$

$$= C \log 6N \int h^2 = C \log 6N \|h\|^2$$

as desired.

To finish the proof, we have to analyze the intersections of the $\tilde{R}_{ij}$; we'll number rows by j and columns by i. All we know about the precise position of $\tilde{R}_{ij}$ is that it intersects Q_{ij}. The largest possible intersection of two $\tilde{R}_{ij}$ occurs when the two are parallel; for then the measure is like $a^2 N$. If the rectangles are perpendicular, the intersection is like a^2.

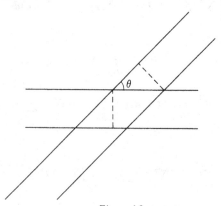

Figure 15

To control the intersection, we maximize it by translating the $\tilde{R}_{ij}$ so that the angle between two is minimized, given the constraint that the two intersect at all.

136

We also dominate any intersection by a maximal intersection; the intersection shown in Figure 15. With this proceedure, the intersection is independent of sliding the rectangles left or right, and since the columns are indexed by i, the estimates we make are independent of i. For convenience, then, we assume $i = i'$.

The area of the intersection in Figure 15 is easily computed; each of the dotted lines is the thickness of one of the rectangles, that is, a, because the height of the parallelogram is a. The width of the parallelogram satisfies $\sin\theta = \frac{a}{N}$, and

$$\left|\tilde{R}_{ij} \cap \tilde{R}_{i'j'}\right| = \frac{a^2}{\sin\theta}$$

(This is inaccurate if $\theta = 0$; the correction is done below).

The angle θ is easily computed, given the conditions $\tilde{R}_{ij} \cap \tilde{R}_{i'j'} \neq \phi$; $\tilde{R}_{ij} \cap Q_{i'j'} \neq \phi$. To maximize the intersections of the $\tilde{R}_{ij}$, the angle θ must be minimized. The smallest possible θ to take is when R_{ij}, $R_{i'j'}$ just touch each other, as in Figure 16.

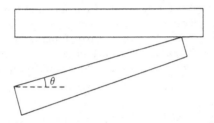

Figure 16

Then

$$\sin\theta = \cos(\frac{\pi}{2} - \theta) \approx \frac{a|j - j'|}{\delta N} = \frac{j - j'|}{N}.$$

Then

$$\left|\tilde{R}_{ij} \cap \tilde{R}_{i'j'}\right| \leq \frac{a^2 N}{|j - j'|}.$$

If $j = j'$, we can get a better estimate, so in general,

$$\left|\tilde{R}_{ij} \cap \tilde{R}_{i'j'}\right| \leq \frac{a^2 N}{|j - j'| + 1}.$$

8.7 REMARK: The above estimate on $\mathcal{C}$ is best possible; we now prove this fact. We choose $a = 1$; let $f(x) = (1 + |x|)^{-1} if |x| \leq N$; let $f = 0$ otherwise. Then $\|f\|_2 \approx (\log N)^{\frac{1}{2}}$. To estimate $\mathcal{C}f$, assume first that $|x| \leq N$. Then we take R_x to have long direction $\frac{x}{|x|}$, and choose it so that the small side of R_x contains x. Then $|R_x|^{-1} \int_{R_x} f$ is radial, and we evaluate it for $x = (|x|, 0)$. Then

$$\frac{1}{N} \int_{-1}^{1} \int_{|x|-N}^{|x|} \frac{1}{1 + |y|} dy_1 dy_2$$

$$\geq CN^{-1} \int_0^{|x|} \frac{1}{1 + y_1} dy_1 \geq CN^{-1} \log|x|.$$

Thus $\|\mathcal{C}f\|_2^2 \geq CN^{-2} \int_0^N \log^2 r \, dr \approx C \log^2 N$.

137

SECTION 8.3 THE CORDOBA MULTIPLIER THEOREM

8.8 NOTATION:

Let ϕ be a smooth function on $\mathbb{R}^1$ with $\phi(x) = 0 \; for \; |x| > 1$. Define ϕ_δ on $\mathbb{R}^2$ by

$$\phi_\delta(\xi) = \phi\left(\delta^{-1}(|\xi| - 1)\right).$$

Let S_δ denote the operator with multiplier ϕ_δ.

8.9 THEOREM.

$$_4\|S_\delta\| \leq C |\log \delta|^{\frac{1}{4}}.$$

PROOF:

The patterns of previous multiplier theorems–say the Marcinkiewicz theorem–suggest that we need isolation of the different chunks of ϕ_δ, and then control of the chunks by a maximal function, and then control of the maximal function. A lot of pieces are already in place; for example the control of the maximal function. We start with defining the pieces.

Intuitively we think of ϕ_δ as the characteristic function of an annulus of thickness δ. The chunks we expect to isolate from each other are the different directions. The function in Theorem 8.2 will do the isolation, but it will not take us from $S_\delta f$ to $\left(\sum |S_j S_\delta f|^2\right)^{\frac{1}{2}}$; instead, it takes us from $\sum |S_j f|^2$ to f. This means we have to do our own isolating of directions.

The computations we did in 8.4 suggest that the curvature of the circle will provide isolation if we choose rectangles of size $\delta \times \delta^{\frac{1}{2}}$. We need to do the decomposition smoothly; taking a smooth function ϕ on S^1; we assume $\phi(\theta) = 0$ if $|\theta| > \frac{1}{2}$, and

$$\phi_j(\theta) = \phi\left(\delta^{-\frac{1}{2}}\left(\theta - 2\pi j[\delta^{\frac{1}{2}}]\right)\right).$$

We also define $\mu_j(\xi) = \mu_j(r, \theta) = \phi_j(\theta)\phi_\delta(r)$, and we let T_j denote the corresponding operator. Intuitively, μ_j is the characteristic function of a rectangle of size $\delta \times \delta^{\frac{1}{2}}$ centered at $j[\delta^{\frac{1}{2}}]$ As in the discussion in 8.4, to avoid interference from different directions, we need to restrict our angles to chunks of the circle; this has the effect of multiplying all final estimates by a factor of 8. We may then assume that $0 \leq j \leq [\delta^{-\frac{1}{2}}]/8$. These are our first decompositions.

To begin the estimates:

$$\|S_\delta f\|_4^4 = \|\sum T_j f\|_4^4 = \int \left|\left(\sum T_j f\right)^2\right|^2$$

$$= \int \left|\sum T_j f \sum T_k f\right|^2 = \int \left|\sum \widehat{T_j f} \star \sum \widehat{T_k f}\right|^2$$

$$= \int \left|\sum_{jk} \widehat{T_j f} \star \widehat{T_k f}\right|^2.$$

138

Since each $x \in \mathbb{R}^2$ belongs to at most 9 of the supports of $\widehat{T_j f} \star \widehat{T_k f} = \mu_j \hat{f} \star \mu_k \hat{f}$, and this is independent of δ, the last integral is bounded by

$$C \int \sum \left| \widehat{T_j f} \star \widehat{T_k f} \right|^2$$

$$= C \int \sum |T_j f T_k f|^2 = C \int \sum |T_j f|^2 \, |T_k f|^2$$

$$= C \int \left(\sum |T_j f|^2 \right)^2 = C \left\| \left(\sum |T_j f|^2 \right)^{\frac{1}{2}} \right\|_4^4 .$$

We get our first decomposition:

$$\| S_\delta f \|_4 \leq C \left\| \left(\sum |T_j f|^2 \right)^{\frac{1}{2}} \right\|_4 .$$

Now its the job of the g-function to get us back from $\sum |T_j f|^2$ to f. If we let I_j denote the projection of the support of ϕ_j onto the ξ_2 axis, we certainly get $\chi_{I_j} \mu_j = \mu_j$. But Remark 8.3 puts some requirements on admissible I_j: they have to be disjoint, bounded in length above and below, and the distance between them has to be bounded below. As it stands, we took ϕ_j smooth, so our I_j overlap. This is easily overcome by taking odd-numbered and even-numbered pieces separately, multiplying all our final estimates by 2. We will not even bother to change notation. Then the I_j are disjoint, have length $C\delta^{\frac{1}{2}} \sin(\frac{\pi}{2} - \theta_j)$; $\theta_j = 2\pi j \delta^{\frac{1}{2}}$. Since we have cleverly restricted $0 \leq \theta_j \leq \frac{\pi}{4}$, the lengths are all bounded by $C\delta^{\frac{1}{2}}$. Of course the separation between the I_j is comparable to the missing pieces, which is again comparable to $C\delta^{\frac{1}{2}}$. Thus we can use 8.2 to get the estimate

$$\left\| \left(|S_j f|^2 \right)^{\frac{1}{2}} \right\|_4 \leq C \| f \|_4$$

if we ever get that far.

The next step in the proceedure is the control of $T_j f$ by a maximal function. Now $T_j f = \check{\mu}_j \star f$, where intuitively, $\check{\mu}_j \star f$ just averages f over a rectangle of size $\delta^{-1} \times \delta^{-\frac{1}{2}}$. Let this rectangle be denoted by R_j. Then $T_j f \approx \frac{1}{|R_j|} \int_{R_j} f$, and

$$\int |T_j f|^2 g = \int \left| \int |R_j|^{-1} \chi_{R_j} (x - y) f(y) dy \right|^2 g(x) dx$$

$$\leq \int \left(\int \chi_{R_j}^2 (x - y) dy \right) \left(|R_j|^{-2} \chi_{R_j}^2 (x - y) |f(y)|^2 dy \right) g(x) dx$$

$$= \int |R_j| |R_j|^{-2} \int \chi_{R_j} (x - y) g(x) |f(y)|^2 dy dx = \int |f(y)|^2 |R_j|^{-1} \chi_{R_j} \star g.$$

Therefore,

$$\int \sum |T_j f|^2 g = \int \sum |T_j S_j f|^2 g \leq \sum \int |S_j f|^2 |R_j|^{-1} \chi_{R_j} \star g.$$

139

If we are going to follow the proof of the Marcinkiewicz theorem, we have to make $|R_j|^{-1}\chi_{R_j} \star g$ independent of j. We do this by taking the supremum over all j; this is dominated by a supremum over all directions, with rectangles of size $\delta^{-1} \times \delta^{-\frac{1}{2}}$. This is a Cordoba maximal function, with $a = \delta^{-\frac{1}{2}}, N = \delta^{-\frac{1}{2}}$. Then

$$\int \sum |T_j f|^2 g \leq \int \sum |S_j f|^2 \mathcal{C}(g)$$

$$\leq \left(\int \left(\sum |S_j f|^2 \right)^2 \right)^{\frac{1}{2}} \|\mathcal{C} f\|_2$$

$$\leq \left\| \left(\sum |S_j f|^2 \right)^{\frac{1}{2}} \right\|_4^2 \|\mathcal{C} f\|_2$$

$$\leq C \|f\|_4^2 \left(\log \delta^{-\frac{1}{2}} \right)^{\frac{1}{2}} \|g\|_2.$$

Taking the supremum over all g with $\|g\|_2 = 1$;

$$\left\| \left(\sum |T_j f|^2 \right)^{\frac{1}{2}} \right\|_4^2 \leq C |\log \delta|^{\frac{1}{2}} \|f\|_4^2,$$

and this proves the theorem.

Now, there are a lot of technical details. We are not convolving with the characteristic function of a rectangle. On the other hand, the convolution kernel is in $\mathcal{S}$, so we can estimate it using Schwartz function ideas. First, notice that the μ_j are all rotates of one function, and we may estimate the convolution kernels in terms of one, K_0. Intuitively,

$$K_0(x_1, x_2) = \delta^{\frac{3}{2}} \check{\phi} \left(\delta x_1, \delta^{\frac{1}{2}} x_2 \right),$$

and using the idea that $\phi \in \mathcal{S}$,

$$|K_0(x_1, x_2)| \leq C \delta^{\frac{3}{2}} |\delta x_1|^{-p} |\delta^{\frac{1}{2}} x_2|^{-q}$$

for all p, $q \geq 0$. This means that K_0 decays so rapidly that we can dominate it by the sum of characteristic functions of rectangles. Let

$$R_i^0 = \{(x_1, x_2)| \ |x_1| < 2^i \delta^{-1}, \ |x_2| < 2^i \delta^{-\frac{1}{2}}\};$$

we claim that

$$|K_0(x)| \leq C \sum 2^{-i} |R_i^0|^{-1} \chi_{R_i^0}(x).$$

To prove this, pick $x \in R_{k+1}$; we assume x is not in R_k. If $|x_1| > 2^k \delta^{-1}$,

$$|K_0(x)| \leq C \delta^{\frac{3}{2}} |\delta x_1|^{-3} |\delta x_2|^0$$

$$\leq C \delta^{\frac{3}{2}} |\delta 2^k \delta^{-1}|^{-3} = C \delta^{\frac{3}{2}} 2^{-3k}$$

140

whereas

$$\sum 2^{-i} |R_i^0|^{-1} \chi_{R_i^0}(x) = \sum_{k+1}^{\infty} 2^{-i} \delta^{\frac{3}{2}} 2^{-2i}$$

$$= C \delta^{\frac{3}{2}} 2^{-3k}.$$

The rest of the proof is the same as before;

$$|T_j S_j f| \leq \sum 2^{-i} \left| |R_i^j|^{-1} \chi_{R_i^j} \star S_j f \right|$$

and

$$\left\| \left(\sum |T_j S_j f|^2 \right)^{\frac{1}{2}} \right\|_4$$

$$\leq \sum 2^{-i} \left\| \left(\sum_j \left| |R_i^j|^{-1} \chi_{R_i^j} \star S_j f \right|^2 \right)^{\frac{1}{2}} \right\|_4.$$

Since the 2^{-i} converge if left alone, it is enough to prove that the L^4 norm is dominated independent of i. This can be done exactly as before, except that the rectangles R_i^j have dimension $2^i \delta^{-1} \times 2^i \delta^{-\frac{1}{2}}$. This means we take the Cordoba maximal function with $a = 2^i \delta^{-\frac{1}{2}}$, $N = \delta^{-\frac{1}{2}}$. The norm we get, $\log N$, is independent of i.

This at last completes the proof, except for the technical detail that μ_0 is not really a dilate of a square, $\psi \left(\delta^{-1} \xi_1, \delta^{-\frac{1}{2}} \xi_2 \right)$. The real dilations are in the (r, θ) directions, and this does not mix so well with Fourier transforms. We now show how to overcome this problem. Our intuitive estimates are based on the principle that the Fourier transform of a rectangle is again a rectangle; the precise version of this looks at derivatives of Schwartz functions. First of all,

$$|x_1 K_0| \leq \|x_1 K_0\|_\infty \leq C \|D_{\xi_1} \mu_0\|_1.$$

We will show that

$$\|D_{\xi_1} \mu_0\|_1 \leq C \delta^{\frac{1}{2}},$$

and this implies that

$$|K_0(x_1, x_2)| \leq C \delta^{\frac{3}{2}} |\delta x_1|^{-1},$$

which is typical of the results we need.

First,

$$D_{\xi_1} \mu = \frac{\partial r}{\partial \xi_1} \frac{\partial \mu}{\partial r} + \frac{\partial \theta}{\partial \xi_1} \frac{\partial \mu}{\partial \theta}$$

$$= \frac{\xi_1}{(\xi_1^2 + \xi_2^2)^{\frac{1}{2}}} \frac{\partial}{\partial r} \phi_\delta(r) \phi(\theta)$$

$$- \frac{\xi_2}{(\xi_1^2 + \xi_2^2)^{\frac{1}{2}}} \phi_\delta(r) \frac{\partial}{\partial \theta} \phi(\theta).$$

141

For $\xi \in support \; \mu_0$, $|\xi_1| \leq 2$; $\xi_1^2 + \xi_2^2 \geq \frac{1}{2}$, $|\xi_2| \leq 2$, and certainly

$$\frac{\partial \phi_\delta}{\partial r} = \delta^{-1} \phi' \left(\delta^{-1}(r - 1) \right)$$

$$\frac{\partial \phi}{\partial \theta} = \delta^{-\frac{1}{2}} \phi' \left(\delta^{-\frac{1}{2}} \theta \right).$$

Thus,

$$|D_{\xi_1} \mu_0| \leq C\delta^{-1} \left(|\phi' \psi| + \delta^{\frac{1}{2}} |\phi \psi'| \right) \leq C\delta^{-1}.$$

Finally,

$$\|D_{\xi_1} \mu_0\|_1 \leq C\delta^{-1} |support \; \mu_0| = C\delta^{-1} \delta^{\frac{3}{2}}.$$

This and similar estimates, complete the proof of the Theorem.

8.10 COROLLARY. *Bochner-Riesz means are bounded on the optimal range.*

PROOF: We start with a partition of unity $1 = \sum \phi_i$. We assume that the ϕ are smooth, radial, supported radially in $[1 - 2^{-i}, 1 - 2^{-i-2}]$ and $|D_r^p \phi_i| \leq C2^{pi}$. Let $\mu_i = \phi_i \mu_\lambda$; note that $\mu_\lambda = \sum \mu_i$.

Intuitively, μ_i is approximately ϕ_i times the largest value of μ_λ on the support of ϕ_i, which is $2^{-\lambda i}$. The L^4 norm of μ_i therefore ought to be like $2^{-\lambda i} \left| \log 2^{-i} \right|^{\frac{1}{4}}$. If $\lambda > 0$, this sums to a finite amount, so that all the Bochner-Riesz operators T_λ are bounded on L^4 as long as $\lambda > 0$. Interpolation now provides the rest of the optimal range.

Of course this is only intuition; to give a precise proof, we re-examine the proof of 8.9. We notice that μ_i enjoys all the properties of the ϕ_δ used there, with $\delta = 2^{-i}$. The only properties we need to verify before simply copying that proof are the estimates on $|K_0|$. What we shall show is that

$$|K_0(x_1, x_2)| \leq C\delta^\lambda \delta^{\frac{3}{2}} |x_1 \delta|^{-p} |x_2 \delta^{\frac{1}{2}}|^{-q}$$

and, as in 8.9, this gives an L^4 norm of $\delta^\lambda \leq \delta^{\frac{1}{4}}$, which is what we needed.

The estimates on K_0 follow the same pattern as those in 8.9. A typical estimate would require that

$$\left| \frac{\partial}{\partial r} \mu_i \right| \leq \delta^\lambda \delta^{-1}.$$

But

$$\left| \frac{\partial}{\partial r} \mu_i \right| \leq \left| \frac{\partial}{\partial r} \phi_i \mu_\lambda \right| + \left| \frac{\partial}{\partial r} \mu_\lambda \phi_i \right|.$$

Now $\left| \frac{\partial}{\partial r} \phi_i \right| \leq C2^i$, while if $r \in support \; \phi_i$, $|\mu_\lambda(r)| \leq \delta^\lambda$. Similarly,

$$\left| \frac{\partial}{\partial r} \mu_\lambda \right| \leq C \left| (1 - r^2)^{\lambda - 1} \right|,$$

and for $r \in support \; \phi_i$, $1 - \delta < r < 1 - \frac{\delta}{4}$, so that

$$\left| \frac{\partial}{\partial r} \mu_\lambda \right| \leq C\delta^{\lambda - 1}.$$

The rest of the estimates follow as in the proof of Theorem 8.9

In this section we shall prove the L^p, L^q boundedness of restriction in $\mathbb{R}^2$ in the optimal range. We begin with the standard model; we look at an annulus instead of the unit circle. Note that

$$\int_{S^1} |\hat{f}(\theta)|^q d\theta$$

$$\lim_{\delta \to 0} \frac{1}{2\delta} \int_{1-\delta}^{1+\delta} \int_{S^1} |\hat{f}(r,\theta)|^q r dr d\theta,$$

so that we can recover restriction from annuli. Let $0 \le \phi \le 1$ be as in 8.8, with $\chi_{(1-\delta,1+\delta)}(|\xi|) \le \phi_\delta(\xi)$. It is enough to prove that

$$\frac{1}{2\delta} \int_{1-\delta}^{1+\delta} \int_{S^1} |\phi_\delta(r)\hat{f}(r,\theta)|^q r dr d\theta \le C\|f\|_p,$$

or,

$$\|\phi_\delta \hat{f}\|_q \le \delta^{\frac{1}{q}} \|f\|_p.$$

Let $Rf = \phi_\delta \hat{f}$; then

$$R^* g = \widehat{\phi_\delta g} = \hat{\phi}_\delta \star \hat{g} = S_\delta \hat{g}.$$

8.11 THEOREM. *Assume* $q = \frac{p'}{3}$.

$$\|S_\delta \hat{f}\|_{p'} \le C\delta^{\frac{1}{q}} \|f\|_{q'}; \quad 1 \le p < \frac{4}{3}$$

$$\|S_\delta \hat{f}\|_4 \le C\delta^{\frac{3}{4}} |\log \delta|^{\frac{1}{4}} \|f\|_4.$$

PROOF: If $p \le \frac{4}{3}$, then $r = \frac{p'}{2} \ge 2$, and

$$\|S_\delta \hat{f}\|_{p'}^{p'} = \|S_\delta \hat{f} S_\delta \hat{f}\|_{p'}^r$$

$$\le \|\widehat{S_\delta \hat{f}} \star \widehat{S_\delta \hat{f}}\|_{r'}^r = \|(\phi_\delta f) \star (\phi_\delta f)\|_{r'}^r$$

$$= \|\sum_{j,k}(\phi_\delta \phi_j f) \star (\phi_\delta \phi_k f)\|_{r'}^r$$

$$\le C\left(\sum_{j,k} \int |(\phi_\delta \phi_j f) \star (\phi_\delta \phi_k f)|^{r'}\right)^{\frac{r}{r'}};$$

the last step uses almost orthogonality as in the proof of 8.9. We shall show shortly that if $1 \le r' \le \infty$,

$$\|(\phi_\delta \phi_j f) \star (\phi_\delta \phi_k f)\|_{r'}$$

$$\le C\left[\frac{\delta^{\frac{3}{2}}}{|j-k|+1}\right]^{\frac{1}{r}} \|\phi_j f\|_{r'} \|\phi_k f\|_{r'}.$$

This gives us a bound of

$$C\delta^{\frac{3}{2}}\left(\sum_k \|f\phi_k\|_{r'}^{r'}\left[\sum_j \frac{\|f\phi_j\|_{r'}^{r'}}{(|j-k|+1)^{\frac{r'}{r}}}\right]\right)^{\frac{r}{r'}}.$$

Let $b_j = \|f\phi_j\|_{r'}^{r'}$. Then we need to control

$$C\delta^{\frac{3}{2}}\left(\sum_k b_k\left[\sum_j \frac{b_j}{(|j-k|+1)^{\frac{r'}{r}}}\right]\right)^{\frac{r}{r'}}$$

$$\leq C\delta^{\frac{3}{2}}\left(\sum_k b_k^s\right)^{\frac{r}{r's}}\left(\sum_k\left[\sum_j b_j(|j-k|+1)^{1-r'}\right]^{s'}\right)^{\frac{r}{r's'}}.$$

If $p < \frac{4}{3}$, $2 - r' < 0$ and we choose $\frac{1}{s'} = \frac{1}{2} - (2 - r')$, and we use the theory of discrete fractional integration (Hardy-Littlewood-Polya [26]) to get a bound of

$$C\delta^{\frac{3}{2}}\left(\sum_k b_k^s\right)^{\frac{2r}{r's}}.$$

But $r's = q'$ and we may estimate $\|\phi_j f\|_{r'}^{r's}$ by using Holder's inequality from r' to q'. This gives us

$$C\delta^{\frac{3}{2}}\delta^{\frac{3}{2}(\frac{q'}{r'}-1)\frac{2r}{q'}}\left(\sum\int|\phi_k f|^{q'}\right)^{\frac{2r}{q'}}$$

$$= C\delta^3\|f\|_{q'}^{p'}.$$

This is what we needed.

On the other hand, if $p = \frac{4}{3}$, $r' = r = s = s' = 2$, and we may use the estimate

$$\left(\sum_k\left[\sum_j \frac{b_j}{|j-k|+1}\right]^2\right)^{\frac{1}{2}}$$

$$\leq \left(\sum(|j-k|+1)^{-1}\right)\left(\sum b_j^2\right)^{\frac{1}{2}}.$$

This gives us a bound

$$C\delta^{\frac{3}{2}}|\log\delta|\sum b_j^2 = C\delta^{\frac{3}{2}}|\log\delta|\sum\|f\phi_j\|_2^4$$

$$\leq C\delta^{\frac{3}{2}}|\log\delta|\|f\|_4^4.$$

To prove the estimate

$$\|(\phi_\delta\phi_j f)\star(\phi_\delta\phi_k f)\|_{r'}$$

144

$$\leq C \left[\frac{\delta^{\frac{3}{2}}}{|j-k|+1} \right]^{\frac{1}{r}} \|\phi_j f\|_{r'} \|\phi_k f\|_{r'},$$

we use a bilinear interpolation. If $B(f,g) = f \star g$ is the bilinear map

$$B : L^p(S_j) \times L^p(S_k) \to L^p(S_j + S_k),$$

we shall interpolate B from $p = 1$ and $p = \infty$. Then

$$\|(\phi_\delta \phi_j f) \star (\phi_\delta \phi_k f)\|_1$$
$$\leq \|\phi_\delta \phi_j f\|_1 \|\phi_\delta \phi_k f\|_1$$
$$\leq \|\phi_\delta\|_\infty^2 \|\phi_j f\|_1 \|\phi_k f\|_1 = \|\phi_j f\|_1 \|\phi_k f\|_1.$$

In the case of $p = \infty$,

$$\|(\phi_\delta \phi_j f) \star (\phi_\delta \phi_k f)\|_\infty$$
$$\leq \|\phi_\delta\|_\infty^2 \|\phi_j f\|_\infty \|\phi_k f\|_\infty \|\phi_j \star \phi_k\|_\infty.$$

The usual "area of a parallelogram" trick give us

$$\|\phi_j \star \phi_k\|_\infty \leq C \frac{\delta^{\frac{3}{2}}}{|j-k|+1}$$

and now interpolation proves the result.

8.12 REMARKS: a) Restriction theorems are already known to fail at $p = \frac{4}{3}$; the presence of the $|\log \delta|$ factor explains the failure.

b) It is easy to show that this estimate is optimal; let $S_j = support\ \phi_j$ and let $g \equiv 1$. Then $\|g\|_4^4 = \delta$, and

$$\|S_\delta \hat{g}\|_4^4 = \int \left| \sum \chi_{S_j} \star \chi_{S_k} \right|^2$$
$$\geq \sum \int \left| \chi_{S_j} \star \chi_{S_k} \right|^2.$$

But $\chi_{S_j} \star \chi_{S_k} \geq C \frac{\delta^{\frac{3}{2}}}{|j-k|+1}$ on at least half the support of $\chi_{S_j} \star \chi_{S_k}$, which has measure at least $C\delta^{\frac{3}{2}}(|j-k|+1)$. Thus

$$\|g\|_4^4 \geq C\delta^3 \delta^{\frac{3}{2}} \sum (|j-k|+1)^{-1}$$
$$= C\delta^3 \delta^{\frac{3}{2}} \delta^{-\frac{1}{2}} |\log \delta|.$$

c) These computations are related to our intuition that restriction must fail unless $\widehat{d\sigma} \in L^p$. As above, let $\hat{g} \equiv 1$; then $S_\delta \hat{g} = \hat{\phi}_\delta$, and the above estimate shows that

$$\|\hat{\phi}_\delta\|_4 \approx C\delta^{\frac{3}{4}} |\log \delta|^{\frac{1}{4}}$$

and this estimate is optimal. Since $d\sigma = \lim_{\delta \to 0} \delta^{-1} \phi_\delta$, we again see the failure of $\widehat{d\sigma}$ to be in L^4. This can also be related to curvature properties; we needed to know that no ξ is in more than 9 of the $S_j + S_k$, and that

$$|\chi_{S_j} \star \chi_{S_k}| \leq \frac{|S_j|}{|j-k|+1}.$$

Both of these properties fail miserably for straight lines. For example, let $S = [-1,1] \times \{0\}$; $D_j = \left[j\delta^{\frac{1}{2}}, (j+1)\delta^{\frac{1}{2}} \right] \times [-\delta, \delta]$, $-\delta^{-\frac{1}{2}} \leq j \leq \delta^{-\frac{1}{2}}$. Then the origin is in all of the $D_j + D_k$, so the overlap is enormous. Similarly, $\|\chi_{D_j} \star \chi_{D_k}\|_\infty = |D_j|$, so we get no decrease in $j - k$.

Section 8.5 Remarks and Extensions

THEOREM 8.1): This result is taken from Cordoba [1], though Cordoba remarks it was known earlier. It has recently been proved that the requirements on spacing and uniformity can be completely removed; the result is true for arbitrary disjoint intervals.

REMARK8.4): The almost orthogonality first appreared in C. Fefferman [23]; a weaker version was used in a restriction theorem of C. Fefferman and E. M. Stein [22]. In the latter, one uses the fact that to each length and direction there correspond at most two chords of a circle. A point in an $S_j + S_k$ corresponds to a direction and length in a slightly thickened circle.

THEOREM 8.6): The proof here is taken from Cordoba [9]. The idea of controlling maximal functions through adjoints was first developed by Cordoba; ıc. f. the Appendix to [18]. The original idea goes back to the Kolmogorov-Seliverstov-Plessner theorem; ıc. f. Zygmund [57].

See Wainger [54] for other proofs of the boundedness of these maximal functions.

THEOREM 8.9): The proof here is from Cordoba [10]. Control of T_δ was first established by Cordoba [9], although in this case it was necessary to first control a much stronger maximal function, the "Kayeya maximal function". This takes averages over all rectangles with ratio of sides N. These ideas go back to C. Fefferman's paper [23].

COROLLARY 8.10): This result was first proved by L. Carleson and P. Sjolin [5]. The proof proceedes by decomposing the convolution kernel into pieces, as in 6.9; one then follows the intuition of 6.1 in a direct manner. Instead of showing that the pieces of T_λ are like restriction operators, Carleson and Sjolin prove Plancherel-type and then L^4 type estimates.

A second proof of 8.10 was given by C. Fefferman [23]; historically this came after the introduction of the Kakeya set and was the first paper to integrate Kakeya-set ideas into positive results.

THEOREM 8.11): The first proof of this result when $p \neq \frac{4}{3}$ is due to C. Fefferman and E. M. Stein [23]; Cordoba [11] gave a geometric proof. The result from annuli is a simplified version of the proof in [52].

REFERENCES

1. Bochner, Salomon, *Summation of multiple Fourier series by spherical means*, Transactions AMS. **40** (1936), p. 175-207.
2. Burkholder, D. L., *Harmonic Analysis and Probability*, in "Harmonic Analysis and Probability", Mathematical Association of America.
3. Calderon, A. P., and Zygmund, A., *On the existence of certain singular integrals*, Acta Math. **88** (1952), p. 85-139.
4. Carberry, A., *Pointwise convergence of spherical means*, PhD Thesis, UCLA (1981).
5. Christ, M., *Restriction to hypersurfaces*, PhD Thesis, University of Chicago (1981).
7. Coifman, R. and Fefferman, C., *Weighted norm inequalities for maximal functions and singular integrals*, Studia Math. **51** (1974), p. 241-250.
8. Coifman, R. and Rochberg, R., *Another characterization of BMO*, Proceedings A. M. S..
9. Cordoba, A., *The Kakeya maximal function and the spherical summation multipliers*, Am. J. Math. **99** (1977), p. 1-22.
10. _____, *A note on Bochner-Riesz operators*, Duke J. Math. **46** (1979), p. 505-511.
11. _____, *A note on $\mathcal{F}(L^p)$*, in "Springer Lecture Notes in Math, 779 .".
12. _____, *The multiplier problem for the polygon*, Annals of Math. **105** (1977), p. 581-588.
13. _____, *Geometric Fourier Analysis*, Annales Inst. Fourier **32** (1982), p. 215-226.
14. Cordoba, A. and Fefferman, C., *A weighted norm inequality for singular integrals*, Studia Math. **51** (1974), p. 241-250.
15. Cordoba, A. and Fefferman, R., *On the equivalence between the boundedness of certain classes of maximal and multiplier operators in Fourier analysis*, Proc. Nat. Acad. Sciences **74** (1977), p. 423-425.
16. Cotlar, M., *A combinatorial inequality and its application to L^2 spaces*, Revista Math. Cuyena **1** (1955), p. 41-55.
17. Cunningham, F., *The Kakeya problem for simply connected and for star shaped sets*, American Math. Monthly **78** (1971), p. 114-129.
18. de Guzman, M., "Differentiation of Integrals in $\mathbb{R}^n$", Springer Verlag, Berlin, 1981.
19. de Leeuw, K., *On L^p multipliers*, Annals of Math. **91** (1965), p. 364-379.
20. Fefferman, C., *The multiplier problem for the ball*, Annals of Math. **99** (1971), p. 330-336.
21. _____, *Pointwise convergence of Fourier series*, Annals of Math. **98** (1973), p. 551-571.
22. _____, *Inequalities for strongly singular convolution operators*, Acta Math. **124** (1970), p. 9-36.
23. _____, *A note on spherical summation operators*, Israel J. Math. **15** (1973), p. 44-52.
24. Gilbert, J. E., *Nikisin-Stein theory and factorization with applications*, Symp. Pure Math. **35**, p. 233-267.
25. Hardy, G. H. and Littlewood, J. E., *A maximal theorem with function-theoretic applications*, Acta Math. **54** (1930), p. 81-116.
26. Hardy, G. H., Littlewood, J. E. and Polya, G., "Inequalities", Cambridge University Press, Cambridge, 1934.
27. Herz, C. S., *On the mean inversion of Fourier and Hankel transforms*, Proc. Nat. Acad. Sciences **40** (1954), p. 996-999.
28. Hormander, L., *Estimates for translation invariant operators on L^p spaces*, Acta Math. **104** (1966), p. 249-275.
29. Hunt, R., *On $L(p, q)$ spaces*, L'Ens. Math. **12** (1966), p. 249-275.
30. Jodeit, M., *A note on Fourier multipliers*, Proceedings A.M.S. **27** (1971), p. 423-424.
31. Jones, P., *On the factorization of A_p weights*, Annals of Math. **111** (1980), p. 511-530.
32. Kenig, C. and Tomas, P.A., *L^p behaviour of certain second order partial differentiation operators*, Transaction A.M.S. **262** (1980), p. 521-531.
33. Knapp, A. W. and Stein, E. M., *Intertwining operators for semisimple groups*, Annals of Math. **93** (1971), p. 489-578.
34. Kolmogorov, A.N., *Sur les fonctions harmoniques conjuguees et les series de Fourier*, Fund. Math. **7** (1925), p. 23-28.

35. Paley, R.E.A.C., *A remarkable system of orthogonal functions*, Proceedings London Math. Soc. **34** (1932), p. 241-279.

36. Marcinkiewicz, J., *Sur l'interpolation d'operations*, Comptes Rendus Acad. Sci. Paris **208** (1939), p. 1272-1273.

37. _____, *Sur les multiplicateurs des series de Fourier*, Studia Math. **8** (1939), p. 78-91.

38. Mihilin, S.G., *On the multipliers of Fourier integrals*, Dokl. Akad. Nauk. **109** (1956), p. 701-703.

39. Muckenhoupt, B., *Weighted norm inequalities for the Hardy-Littlewood maximal function*, Transaction A.M.S. **15** (1972), p. 207-226.

40. _____, *Weighted norm inequalities for classical operators*, Proc. Symp. Pure Math. **35** (1979).

41. Riesz, M., *Sur les fonctions conjugees*, Math. Zeit. **27** (1927), p. 218-244.

42. Rudin, W., "Real and Complex Analysis", McGraw Hill, New York, 1966.

43. _____, "Functional Analysis", McGrawHill, New York, 1973.

44. Shapiro, H.S., *Lebesgue constants for spherical partial sums*, J. Approx. Theory **13** (1975), p. 40-44.

45. Stein, E.M., *Interpolation of linear operators*, Transactions A.M.S. **83** (1956), p. 482-492.

46. _____, *On limits of sequences of operators*, Annals of Math. **74** (1961), p. 140-170.

47. _____, *Localization and summability of multiple Fourier series*, Acta Math. **100** (1958), p. 93-147.

48. _____, *The development of the square function in the work of A. Zygmund*, Bulletin A.M.S. **7** (1982), p. 359-376.

49. _____, "Singular Integrals and Differentiability Properties of Functions", Princeton University Press, Princeton, 1970.

50. Stein, E.M. and Weiss, G., "Introduction to Fourier Analysis on Euclidean Spaces", Princeton University Press, Princeton, 1971.

51. Tomas, P.A., *A restriction theorem for the Fourier transform*, Bulletin A.M.S. **81** (1975), p. 477-478.

52. _____, *A note on restriction*, Indiana J. Math. **29** (1980), p. 287-292.

53. Wainger, S., *Special Trigonometric Series in k dimensions*, Memoirs AMS **59** (1965).

54. _____, *Applications of Fourier transforms to averages over lower dimensional sets*, Proc. Symposia Pure Math. **35**, p. 85-94.

55. Watson, G.N., "A Treatise on the Theory of Bessel Functions", Cambridge University Press, Cambridge, 1966.

56. Whitney, H., *Analytic extensions of differentiable functions defined in closed sets*, Transactions A.M.S. **36** (1934), p. 63-89.

57. Zygmund, A., "Trigonometric Series", Cambridge University Press, Cambridge, 1935.

INDEX

Printed in the United States
By Bookmasters